Sustainable Transport Fuels Business Briefing

David Thorpe

News editor, *Energy and Environmental Management*

First published in 2012 by Dō Sustainability
87 Lonsdale Road, Oxford OX2 7ET, UK

ISBN 978-1-909293-10-6 (eBook-ePub)
ISBN 978-1-909293-11-3 (eBook-PDF)
ISBN 978-1-909293-09-0 (Paperback)

A catalogue record for this title is available from the British Library.

At Dō Sustainability we strive to minimize our environmental impacts and carbon footprint through reducing waste, recycling and offsetting our CO_2 emissions, including those created through publication of this book. For more information on our environmental policy see **www.dosustainability.com**.

Page design and typesetting by Alison Rayner
Cover by Becky Chilcott

For further information on Dō Sustainability, visit our website: **www.dosustainability.com**

DōShorts

Dō Sustainability is the publisher of **DōShorts**: short, high-value ebooks that distil sustainability best practice for busy professionals. Each DōShort addresses one sustainability challenge at a time and can be read in 90 minutes.

Be the first to hear about new DōShorts

Our monthly newsletter includes links to all new and forthcoming DōShorts. We also link to a free extract from each new DōShort in our newsletter, and to blogs by our expert authors. Sign up for the newsletter at **www.dosustainability.com**.

Recently published and forthcoming titles you'll hear about include:

- *Promoting Sustainable Behaviour: A Practical Guide to What Works*
- *Sustainable Transport Fuels Business Briefing*
- *How to Make your Company a Recognised Sustainability Champion*
- *Product Sustainability*
- *Corporate Climate Change Adaptation*
- *Making the Most of Standards*
- *Sustainability Reporting for Small and Medium-sized Enterprises*
- *The Changing Profile of Corporate Climate Change Risk*
- *Solar Photovoltaics: Business Briefing*
- *Green Jujitsu: The Smart Way to Embed Sustainability into Your Organisation*
- *The First 100 Days on the Job: How to Plan, Prioritise and Build a Sustainable Organisation*

Write for us, or suggest a DōShort

Please visit **www.dosustainability.com** for our full publishing programme. If you don't find what you need, write for us! Or Suggest a DōShort on our website.

We look forward to hearing from you!

....

Contents

CONTENTS

CHAPTER 1

About This Book

THIS BOOK LAYS OUT, for a global business audience, the exciting developments in the world of sustainable transport. Not the vehicles or modes of transport themselves, but their means of propulsion. This is a fast-moving field with many new technologies and players coming and going with bewildering speed. Some observers are putting their money on electric vehicles, others on hybrids being around for many years to come. Some see electric vehicles as a mere stepping stone to hydrogen-powered fuel cell vehicles, already being seen on city streets. The mere mention of biofuels often provokes fierce arguments about their sustainability, yet they, too, are here to stay and will be filling more and more gasoline tanks.

By the time you finish this e-book, you will be able to speak knowledgeably not only about the pros and cons of all these technologies, the programs around the world furthering their development, and the players large and small, but also the catalysts for change, the successful partnerships and innovative business models being employed. You'll also be able to tell if it's a good idea to install an electric vehicle charging point in your property.

1.1 Who it is for?

You are probably involved in business, NGOs or government and interested in knowing the extent of this fascinating field. You might be hoping to decide where to put your money, which horses to back, or, if investing in new fleet, haulage or mobility of any kind, you'll want to know the best and most sustainable target for your investment.

1.2 Catalysts for change

The race to find sustainable ways of powering transportation of all types is forcing the investment of billions of dollars around the world. Many different routes are being pursued, and many of these will end up finding sizeable markets. The catalysts for this race are not hard to find.

Emissions from transportation are the fastest growing source of global greenhouse gas emissions, with emissions expected to increase 300% by 2050, according to the Worldwatch Institute. Furthermore, pollutants from transportation contribute to around 80% of those emissions, causing 1.3 million premature deaths around the world every year according to Michael Replogle and Colin Hughes of the Institute for Transportation & Development Policy (ITDP). This depressing picture is projected to get worse. The International Energy Agency reckons there are 800 million cars on the world's roads today, but you haven't seen anything yet. Without changes to this trajectory, they will reach two to three billion by 2050. However, despite this, the IEA believes that fuel consumption could be halved within 20 years, setting out its thoughts in two reports published in September 2012: *Technology Roadmap: Fuel Economy of Road Vehicles* and *Policy Pathway: Improving the Fuel Economy of Road Vehicles*.

Will we manage to replace completely all fossil fuels used for transportation in the next half-century? Not a chance. Although, to our children in 2050, contemporary vehicles will appear as quaint as vintage cars and traction engines do to us in 2012, some technologies and fuels will still be recognisable. This is because there are limits to the amount of biofuels and of renewable electricity capacity for electric vehicles that can be developed within this timeframe. So, let us peer down this road to transport fuels in 2050.

CHAPTER 2

The Transport Revolution

2.1 **Rio+20**

THE TREND TOWARDS MORE SUSTAINABLE TRANSPORT is global. The largest financial commitment made at the Rio+20 Conference on Sustainable Development in June 2012 was a pledge by the eight biggest multilateral development banks (MDBs) to commit 500 staff and to dedicate $175 billion for more sustainable transportation in the coming decade. This unprecedented agreement was facilitated by the Partnership on Sustainable Low Carbon Transport (SLoCaT), which brings together 68 MDBs, civil society organisations, UN agencies, and research and industry organisations. But besides climate change and air pollution, several other pressures are fuelling the changes to come.

2.2 **Consumer pressure**

CO_2 emissions from vehicles have been a factor in consumers' choice of vehicles and fuels for over 10 years. This is because for at least this long, in the UK and elsewhere, emission levels of vehicles have been used to assign reductions in their due Vehicle Excise Duty and Company Car Tax. HM Revenue and Customs figures show that Treasury receipts from company cars' tax and duty have fallen by over £500 million since 2005 as a result of this, and of lower sales. Differential tax rates are applied to encourage the take-up of lower-emitting vehicles, including a zero rating for electric vehicles, introduced in 2009. Since then, manufacturers have been required by law to display fuel consumption and CO_2 data for their cars in promotional material. A voluntary agreement has also been in place with industry for a colour-coded new car efficiency label to be displayed at point of sale.

In the UK, as elsewhere, the government also provides consumers with advice about buying a 'green' car. Most drivers are, however, likely to be more aware of their vehicle's fuel efficiency than anything else. But CO_2 emissions and fuel efficiency are directly correlated, so that as emissions have fallen, fuel efficiency has risen. Drivers can therefore immediately see a direct cost-saving for purchasing low-emission vehicles. The recession and ongoing increases in fuel costs are also helping to focus purchasers' minds on more fuel-efficient vehicles.

This explains why, in 2011, UK registrations of all alternatively fuelled cars rose to a record 25,456 units, or 1.3% share of the total market, according to the Society of Motor Manufacturers and Traders. As a transition vehicle, petrol-electric hybrids accounted for 92% of all 2011

FIGURE 2.2A. Change in new car CO2 and MPG (miles per gallon), 2000–2011.

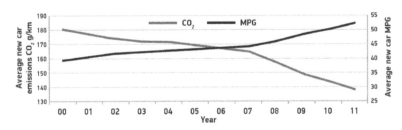

SOURCE: Society of Motor Manufacturers and Traders (SMMT)

FIGURE 2.2B. Registrations of alternatively fuelled cars, 2000–2011.

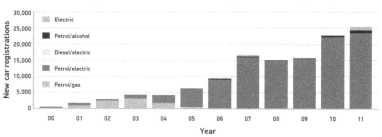

SOURCE: Society of Motor Manufacturers and Traders (SMMT).

alternative fuel vehicle registrations. Their average CO_2 emissions were 103g/km, 25% lower than the UK average.

2.3 Regional legislation

Many countries around the world now have mandates or targets for increasing the proportion of biodiesel or bioethanol in petroleum or diesel. Twenty-two countries have mandatory directions, while six have targets. A further five are planning targets. The result has been a global biofuel production rise from 16 billion litres in 2000 to more than 100 billion litres in 2010. They now provide about 3% of total road transport fuel globally, on an energy basis. In the United States the figure is 4% of road transport fuel.

2.3.1 European Union

Legislation exists both to cut transport emissions and promote biofuels. Cutting emissions may mean making vehicles use conventional fuel more efficiently, or burn different fuels, or employ different technology, such as batteries and fuel cells.

2.3.1.1 Emission reduction

European transport emission targets will cut average emissions from new cars to 95 grams of CO_2 per km (g CO_2/km) by 2020. The car industry is already working towards a mandatory target of 130 g, which is being phased in from 2012 to 2015. Emissions from vans will be reduced to 147 g CO_2/km in 2020, reduced from 181.4 g in 2010, with a mandatory target of 175 g in 2017.

Cars and vans together account for around 15% of EU CO_2 emissions, including emissions from fuel supply. The existing CO_2 regulation for passenger vehicles, which expires in 2015, has already led to impressive results: average new car CO_2 emissions have dropped from about 160 g/km in 2006 to 136 g/km in 2011; a 15% reduction. The annual rate of reduction is about twice what it was before introduction of mandatory emission targets. So far, despite protesting in advance about the setting of targets, manufacturers have developed low CO_2-emitting cars ahead of them, albeit in small numbers. New petrol vehicles can tolerate fuel with a blend of up to 5% biofuel. New diesel vehicles can tolerate a blend up to 7%. Research and development is ongoing to develop cars that can tolerate higher blends of biofuel.

However, manufacturers face other legislative challenges too, chief amongst which is a European standard to reduce regulated emissions such as nitrogen dioxide and particulates. Any changes in the standards, as happened in 2011 with Euro 5 (which regulates the emission of NOx and PM pollutants), results in changes in engine design, but also makes manufacturers think about alternative fuels. Optimising an engine for

CO_2 emission reduction may not always reduce other emissions, and vice versa. A manufacturer may consider that switching to electric vehicles entirely would, potentially, help to meet both of these targets.

2.3.1.2 **Biofuels**

The use of liquid biofuels in transport in the UK and Europe has been driven by the Renewable Transport Fuel Obligation (RTFO). This requires that biofuel penetration reaches 5% by volume (4% by energy) in 2013/14. The proportion of biofuels in the UK transport fuel mix was 3.3% by volume in 2009/10. No specific route is given in the Renewable Transport Fuels Obligation to achieve this target.

The Renewable Energy Directive requires 10% of transport fuel to be renewable by 2020. A rise of 8% is supported by the UK's Committee on Climate Change, and by the important 2008 Gallagher Review, commissioned by the UK government, which recommended a slow-down in the use of biofuels until sustainability criteria were formally established. Advanced biofuels and cellulosic biofuels (see below) must demonstrate that they meet minimum greenhouse gas (GHG) reduction standards of 50% and 60% respectively, based on a lifecycle assessment (including indirect land-use change) in comparison with the petroleum fuels they displace.

In Switzerland, the Federal Act on Mineral Oil mandates a 40% GHG reduction of biofuels in order to qualify for tax benefits.

2.3.2 United States

In the United States, the Renewable Fuel Standard II program, created by a 2007 Act, specifies annual volume requirements for renewable fuels.

In 2012 this is 13.2 billion gallons, rising to 36 billion gallons by 2022. It contains the following percentage standards, which represent the ratio of renewable fuel volume to non-renewable gasoline and diesel volume in 2012:

- Cellulosic biofuel 0.010%

- Biomass-based diesel 0.91%

- Advanced biofuel 1.21%

- Renewable fuel 9.21%

Even without the Standard, a third of the US gasoline supply in the most populous parts of the country must contain ethanol to meet the requirements of the 1990 Clean Air Act. These requirements apply to domestic and foreign producers and importers of renewable fuel used in the US. Bioethanol is also as much as $1 cheaper than other types of octane boosters such as reformate, which refiners use to increase the efficiency of their fuel.

Forty percent of American-grown corn is now sold for ethanol production; 10%, or 850,000 barrels per day, of all gasoline sold there is ethanol. Because of the 2012 drought in America, prices of corn have risen. As a result, a 10% reduction in bioethanol production is envisaged in 2013. However, it is thought by observers to be unlikely that the Standard will be amended by the Administration solely on the basis of one year's weather.

Tax credit schemes and capital allowance exist in the US to support cellulosic biofuels and biodiesel, allowing producers to claim a $1 per gallon production tax credit. Algae-based fuels are included in this, and there is a small agri-biodiesel producer credit of 10 cents per gallon.

To promote next-generation biofuels, the US Department of Agriculture, through its Bioenergy Program for Advanced Biofuels, has made nearly $19.4 million available to 125 advanced biofuel producers growing non-food feedstocks, to support the research, investment and infrastructure needed to build a diverse supply chain.

The US military is helping to promote biofuels too, through its commitment to use 50–50% blends of conventional and biofuels in all of its ships, jets and helicopters, which now burn about 300 million gallons of biofuels per year. Separately, the Air Force is incorporating biofuels into its stores and the Army has set a goal of being petroleum-free by 2040. The policy has been criticized because of its cost but, according to Navy Secretary Mabus, continued military commitment to biofuels will drive down their price, particularly because successful military use will lead to broader commercial use.

2.4 The IEA roadmap

The International Energy Agency's roadmap for reducing greenhouse gas emissions, called the Blue Map Scenario, anticipates that by 2050 biofuels will power the following percentages of different transport modes: 37% of road passenger transport, 26% of road freight transport and aviation, and 11% of shipping. Twenty-seven percent of all transport fuels will be biofuels, 13% electricity and 7% hydrogen. The remaining 53% will be fossil fuels. This will require a total of 32 EJ (1 exajoule = 10^{18} joules). The greatest demand for biofuels will be in OECD countries. China and India will also be significant markets; India aims to meet 20% of its diesel demand with fuel derived from plants by 2017.

2.5 **UK capital allowances**

Businesses claim capital allowances to reduce the tax they pay on profits for the purchase of certain products. Vehicles are included in this, and in the UK the capital allowance has, since April 2002, been designed to encourage the take-up of lower emission vehicles. From April 2013 the main rate of capital allowances for business cars will reduce from 160 g/km to 130 g/km. The threshold above which lease rental restriction applies will also reduce from and to the same levels at the same time.

Pure electric vehicles are zero rated for vehicle excise duty and company car tax. Some local authorities also use carbon dioxide emissions as a basis for differential charging with parking permits. Vehicles with emissions below 100 g/km, which are also Euro 5 emissions-level compliant, get a 100% discount on the London congestion charge.

2.6 **Further institutional support**

In the UK, the transport sector is supported by the Department for Business, Innovation and Skills' Automotive Unit, together with the Department for Transport's Office for Low Emission Vehicles (OLEV) and the Low Carbon Vehicle Partnership (LowCVP). Crucial to fostering UK success is Cenex, the UK's first Centre of Excellence for low carbon and fuel cell technologies.

OLEV manages the Plug-In Vehicle Infrastructure Strategy, including grants of £5000 towards the purchasing of electric cars and vans, pilot schemes for the electric vehicle (EV) re-charging infrastructure, R&D, and support for automotive manufacturing and the UK supply chain.

In Europe, biofuels are supported by the European Biofuels Technology Platform. The European Commission's Strategy for a Sustainable European Bioeconomy, proposed in February 2012, aims to shift the European economy towards greater and more sustainable use of renewable resources and processes (for food, feed, energy and industry).

2.7 Carbon capture and storage (CCS)

The capture of carbon emissions during the generation of bioenergy could be a significant driver of take-up of certain types of biofuel. Carbon dioxide produced from the fermentation or gasification of organic materials to make biofuels is of greater purity, and cheaper and easier to capture than that from flue gases produced in fossil-fuel power plants. CCS projects of the latter type are currently unproven at scale and extremely expensive to implement. By contrast, bioenergy with carbon capture and storage (BECCS) is comparatively cheaper. As with conventional CCS, the gas is injected underground for long-term geological storage. A demonstration project in Illinois is underway. In this future scenario, carbon credits could be claimed to help the financing of such a scheme. This is a third-generation biofuels scenario.

2.8 Flywheel technology

Although not strictly speaking a fuel, it's worth mentioning this technology here. There's nothing new about flywheels as a form of energy storage and their use in regenerative braking. However, they are undergoing serious re-evaluation as solutions for the most pressing problem of storing electricity, whether in vehicles or for the grid. Amongst these is a new technology based on centrifugal energy, which is being investigated by Williams Formula One. Part of a hub of automotive design and engineering around Oxford, Williams, which is commercialising the technology through Williams Hybrid Power, is one of the many innovative start-ups in the area. Many high-tech companies are involved in producing the most sustainable, fastest cars in the world for the sporting scene, thereby pioneering technologies that have spin-offs in consumer vehicles.

2.9 **Compressed air**

Another emerging technology, which is also not a fuel but is nevertheless helping to make internal combustion engines far more fuel efficient, is the air compressor. Several companies are developing variations of these. For example, Aeristech, based in Warwickshire, UK, has produced a 35-kW air compressor that can help a car accelerate to 150,000 rpm in 0.5 seconds. It works by instantaneously compressing exactly the right amount of air into an engine to enhance the combustion of the fuel at any time in the engine cycle. This gives very precise control of rapid acceleration of electric motors and generators.

Like many other serious start-ups in this competitive field, the company has as its business model the licensing of rights to its technology to those in the automotive industry who have found applications for it. Aeristech is also exploring opportunities to supply motor control technology modules to its licensees. This will enable it to secure a revenue stream to support research and development, although it does ensure protection from acquisition, always a possibility in these cases; companies who own a patent on a highly interesting technology are often bought out by OEMs (auto-manufacturers) simply to avoid their rivals getting their hands on the technology.

2.10 **The auto-maker's gamble**

It takes between three and five years for an auto-manufacturer to redesign a product line to incorporate technological developments, so any change to fuel and power-train requirements would take longer to kick in than a simple change of fuel composition that requires no adjustments to the engine. This makes it hard for auto-manufacturers to plan ahead. They need legislative certainty, and have to gamble on what the price of different fuels will be in five years or more to determine where to place their investment now. For example, if battery prices fall quickly and petroleum prices rise, then battery-electric or hybrid vehicles will be attractive. On the other hand, if battery prices stay high, this will favour biofuels. Many are therefore experimenting with a range of alternatives; more choice for the consumer, more risk for the OEM – and higher price tags.

CHAPTER 3

Biofuels

3.1 Generations of biofuel

SOURCES OF BIOFUELS ARE TYPICALLY classed as first, second or third generation. Types of biofuel include biodiesel, bioethanol, syndiesel or renewable diesel, methanol, dimethyl ether (DME), and other fuels and fuel additives. Biomethane and hydrogen are gaseous fuels.

3.1.1 First generation

First-generation sources constitute fuels derived from food crops, or crops grown on former virgin rainforest land, or on otherwise agricultural land, thus displacing food crops. They are the edible parts of sugar, starch and oil plants, and include wheat, maize, soya, palm oil and sugar cane. These are controversial because they have contributed to driving up food prices and to the destruction of richly biodiverse habitats. They are blamed for causing indirect land-use change (ILUC – see below). Bioethanol is produced from sugarcane using fermentation, and biodiesel from vegetable oils like palm or rapeseed.

Under new EU proposals that recognise that they are unsustainable, the use of biofuels made from crops such as rapeseed and wheat would be limited to 5% of total energy consumption in the EU transport sector in 2020. All public subsidies for crop-based biofuels will end after the current legislation expires in 2020, effectively ensuring the decline of a European sector now estimated to be worth 17 billion euros ($21.7 billion) a year.

Crop-based fuel consumption accounted for about 4.5% of total EU transport fuel demand in 2011. To achieve the EU's binding target to source 10% of road transport fuels from renewable sources by the end of the decade, it is therefore likely that the European Commission will support the use of advanced non-land using biofuels made from household waste and algae, i.e. third generation biofuels.

3.1.2 Second generation

Second-generation sources attempt to bypass the problems associated with first-generation biofuels. They include:

1. Cellulosic ethanol, made from inedible parts of crops, such as wheat straw and seed husks, forestry and forest residues.

2. Used vegetable oil, which has often been hydrotreated.

3. 'Dedicated' energy crops: fast-growing trees and grasses with a high lignin content such as miscanthus and willow; and oil crops such as jatropha.

4. Other biomass-to-diesel, which could come from a biorefinery process that starts with the gasification of biomass, and fuels produced via the Fischer-Tropsch process.

5. Bio-butanol and DME.

6. Bio-synthetic gas, also known as Biosyngas or BioSNG. The 'SNG' can stand for 'Substitute Natural Gas' as well as 'synthesis'.

3.1.3 Third generation

Third-generation sources include algae and bacteria grown in special conditions, and novel fuels such as furanics, which are derived from the sugars found in cellulose. None of these are yet at a commercial stage but there are many competing technologies vying for funding to achieve commercial breakthrough.

3.2 **Conversion processes**

To convert the raw material into a usable biofuel, many different chemical processes may be used. They can be broken down into two general types:

1. Mature technology already being used widely to produce biofuels on industrial scales. These include fermentation and transesterification or hydrogenation.

2. Emerging processes, currently subject to research and development or demonstration projects, such as cellulosic ethanol production, Fischer-Tropsch synthesis, and pyrolysis.

TABLE 3.2. The blending characteristics of different biofuels

Biofuel	Blending characteristics
Sugar-based bioethanol	E10–E15 (E25 in Brazil) in conventional gasoline vehicles; E85–E100 in FFV or ethanol vehicles
Starch-based bioethanol	Same as above
Cellulosic-bioethanol	Same as above

Conventional biodiesel (FAME – fatty-acid methyl ester)	Up to B20 in conventional diesel engines
Hydrotreated vegetable oil (HVO)	Fully compatible with existing vehicle and distribution infrastructure
Bioass-to-Liquid (BtL) diesel	Same as above
Algae oil based biodiesel/bio-jet fuel	After hydrotreating: fully compatible with existing vehicle and distribution infrastructure
Biogas (biomethane)	After upgrading, Biogas becomes biomethane, and is fully compatible with natural gas vehicles and fuelling infrastructure
Bio-SNG	Same as above
Bio-butanol	Use in gasoline vehicles in blends up to 85%
Dimethylether	Compatible with LPG infrastructure
Methanol	10–20% blends in gasoline; blend up to 85% in FFVs
Sugar-based diesel/ jet-fuel	Fully compatible with existing vehicle and distribution infrastructure

SOURCE: Technology Roadmap: Biofuels for Transport (IEA, 2011).

Notes: 'E' numbers represent ethanol fuel mixtures describing the percentage of ethanol in the fuel by volume. For example, E85 is 85% anhydrous ethanol and 15% gasoline. Blends of E10 or less are used in more than 20 countries around the world. Similarly, B20 is a mixture of 20% biodiesel with 80% of diesel. FFV = flexible fuel vehicle.

3.3 **Types of biofuels**

3.3.1 Bioethanol

In the sugar-to-ethanol process, sucrose is obtained from sugar crops such as sugarcane, sugar beet and sweet sorghum, and fermented to ethanol, which is then concentrated.

The starch-to-ethanol process requires, in addition to this, the hydrolysis of starch into glucose, and uses more energy than the sugar-to-ethanol route. This means that the process's overall economic and environmental efficiency depends to a great extent on the market value of co-products, such as dried distiller's grains with solubles (DDGS) and fructose.

The cost of bioethanol is currently directly related to feedstock prices, which are volatile. Second-generation bioethanol will be derived from cellulosic lignin (see below) and be less vulnerable to such influence. Efficiency could be improved and costs lowered through the use of more effective amylase enzymes, decreased ethanol concentration costs and the sourcing of markets for co-products.

Bioethanol from sugarcane has the highest yield of all crops sources. Brazil produces one fifth of the world's bioethanol, having pioneered its production from sugarcane since the oil scare of the 1970s. Currently, the high cost of capital and operations limit bio-based materials and chemicals to a few locations in the world where corn and cane are plentiful and cheap, like Brazil and America.

3.3.2 Conventional biodiesel

Biodiesel may be produced from soybean, canola, oil palm or sunflower, animal fats and used cooking oil, which are, again, sensitive to feedstock prices. The conversion process uses methanol or ethanol. Untreated raw oils are sometimes used, but this is not recommended due to the risks of engine damage and gelling of the lubricating oil. Co-products of production, mainly protein meal and glycerine, also help boost the economics of biodiesel.

Commercially available methanol, which is used to make biodiesel, is often made from the methane that exudes from oil and gas wells, and burning it, or a methyl ester derived from it, continues to exacerbate climate change. Absurdly, in the UK, while biofuels are taxed under the Hydrocarbon Oil Duties Act 1979 (HODA), methanol made from methane is not.

3.3.3 Biomethane

Biogas (biomethane, CH_4, usually with some CO_2 and hydrogen sulphide present) can be produced through anaerobic (without oxygen) digestion of feedstocks such as organic waste, animal manure and sewage sludge, or from dedicated crops such as maize, grass and crop wheat. Anaerobic digestion is far more sustainable.

Biogas may be used to generate heat and electricity, or upgraded to biomethane by removing the other gases, and injected into the natural gas grid or used as fuel in natural gas vehicles. It is then chemically identical to 'natural gas', which is classified as fossil fuel, but it does not contribute to global warming because its production captures methane otherwise released into the atmosphere through natural decay, and its use as a fuel supplants the use of fossil fuels.

3.3.4 Bio-synthetic gas (bio-syngas or bio-SNG)

Bio-SNG is processed from biomass using gasification, which produces a mixture of gases. To generate biomethane from bio-SNG, it needs to be cleaned, filtered and processed further, using advanced catalytic and chemical processing techniques; these will ultimately combine the hydrogen and carbon monoxide in the gas to form methane. This process is called 'methanation', and it can produce pipeline-quality biomethane once the metering requirements have been met. Of all second-generation biofuels, methane is the one that allows for the highest energy conversion factors in gasification processes, because it is the simplest hydrocarbon that exists.

The first demonstration production plant has been operating since late 2008 in Gussing, Austria, and construction of a 200-MW plant is planned to start in 2013 in Gothenburg, Sweden. Bio-SNG produced this way can be used just like biomethane generated via anaerobic digestion. It may also be converted into liquid advanced biodiesel, bioethanol or other fuels.

The deployment of natural gas vehicles (NGV) has started to grow rapidly, particularly during the last decade, reaching shares of 25% and more of the total vehicle fleet in countries including Bangladesh, Armenia and Pakistan (IEA, 2010d). These vehicles can also be run on biomethane derived from anaerobic digestion or gasification of biomass.

Biomethane case study

In a trial for Coca-Cola Enterprises (CCE)'s logistics department, Cenex, the UK's first Centre of Excellence for low carbon and fuel cell technologies, compared the emissions, fuel consumption, economics, reliability and operability of a 26-tonne Iveco Stralis vehicle running on biomethane with that of an otherwise identical diesel vehicle. The gas vehicle emitted 85.6% and 97.1% less NO_x and PM emissions respectively than the diesel one, as well as 50.3% less well-to-wheel greenhouse gas emissions. The test involved the use of a temporary filling station; with a more efficient permanent station installed at the depot, the greenhouse gas saving increased to 60.7%. Fuel costs were also 12.8% cheaper.

3.3.5 Cellulosic ethanol

This second-generation bioethanol is produced from ligno-cellulosic feedstocks through biochemical conversion into sugars. These are then fermented to ethanol, following the same conversion steps as conventional biofuels. The first large-scale plants demonstrating this technology are now coming into production.

3.3.6 Hydrotreated vegetable oil (HVO)

This is produced by hydrogenating vegetable oils or animal fats and is also known as hydrogenation-derived renewable diesel (HDRD). The first large-scale plants have been opened in Finland and Singapore, but the process has not yet been fully commercialised. In the US, development is supported by the Department of Energy.

3.3.7 Biomass-to-liquid (BtL) diesel

Also referred to as Fischer-Tropsch diesel, this is produced by a two-step process in which biomass is converted to a bio-SNG that is rich in hydrogen and carbon monoxide. After cleaning, the bio-SNG is catalytically converted through Fischer-Tropsch (FT) synthesis (a sequence of chemical reactions) into a broad range of hydrocarbon liquids, including biodiesel and bio-kerosene.

3.3.8 Pyrolysis

Biofuels can also be produced using pyrolysis, which, like gasification, involves heating biomass with limited oxygen supply. Pyrolysis oil, or other thermochemically derived biomass liquids can be used directly as a fuel, or can be converted to biodiesel. This process is still in demonstration stage. Second-generation plant feedstocks such as miscanthus and other lignocellulosic materials may be used.

3.4 Are biofuels sustainable?

Because of concerns about the sustainability of first-generation biofuels, their take-up has so far been limited. There are also questions about the lifecycle greenhouse gas emissions of all biofuels. These emissions of biofuels can be reduced in five ways:

1. By growing dedicated energy crops that do not require prime agricultural land.

2. Growing crops which require limited use of fertiliser and other inputs.

3. Growing crops which require less intensive or energy-hungry processing.

4. Growing crops which help to increase soil carbon sequestration on marginal land.

5. Developing more integrated production systems for food and fuel, such as through agro-forestry and the greater use of co-products.

All of this is crucial to minimising the impact of travel.

3.4.1 The greenhouse gas balance of biofuels

It is important to note that there is, as yet, no standardised methodology for assessing the greenhouse gas emissions of all biofuels. However, comparisons can still be made:

- The cultivation of corn, wheat and rapeseed oil produces, on average, emissions of around 23–28 g of carbon dioxide per megajoule of biofuel, which represents 28–34% of the emissions from conventional fuels. This is due principally to the use of fertilisers, but also to emissions from the soil.

- Oil palm and sugarcane require a little less fertiliser and contribute about half of these emissions.

- Sugar beet is slightly less again, at 12 g of carbon dioxide per megajoule.

- New types of oil crops such as carmeline and jatropha, although at early stages of development as biofuels, could have even lower cultivation emissions, because they require even less fertiliser.

Emissions from the production and transportation of biofuels can be higher than for cultivation, but may be reduced by altering the energy source used to power the processing of the feedstock. The emissions from transporting biofuels are comparable to those for transporting conventional fuels.

Of all the first-generation biofuels, the most sustainable in terms of cultivation is bioethanol derived from crops grown on marginal or abandoned cropland. Brazilian sugarcane ethanol is 17 times more polluting than this, Brazilian soya bean biodiesel 37 times as polluting, and, at the other extreme, Indonesian palm biodiesel grown on peatland rainforest is a staggering 423 times more polluting. Many indirect consequences of growing biofuels are currently unknown.

An EU analysis in 2010 revealed that biodiesel made from North American soybeans has an indirect carbon footprint of 339.9 kilograms of CO_2-equivalent per megajoule; that is four times higher than standard diesel! Biodiesel from European rapeseed was calculated at an indirect carbon footprint of 150.3 kg of CO_2 per gigajoule. Bioethanol from European sugar beet came in at 100.3 kg. Both are much higher than normal diesel or petrol, which is around 85 kg. Bioethanol from Latin American sugarcane and palm oil from Southeast Asia are, surprisingly, estimated below this, emitting 82.3 kg and 73.6 kg respectively.

Further data from the European Commission (Euractiv), issued at the beginning of 2012, found different results. The figures are again grams of carbon dioxide-equivalent per megajoule of fuel for biofuels produced from different crops:

TABLE 3.4.1A. The global warming potential of biofuels produced from different crops

Fuel	gm CO_2/MJ
Tar sands	107 g
Palm oil	105 g
Soybeans	103 g
Rapeseed (canola)	95 g
Conventional gasoline	87.5 g
Sunflower	86 g
Palm oil with methane capture	83 g
Wheat (process fuel not specified)	64 g
Wheat (as process fuel, natural gas used in CHP)	47 g
Corn (maize)	43 g
Sugarcane	36 g
Sugar beet	34 g

SOURCE: EC (2012).

All were bested by second-generation biofuels, as follows (non-land using include algae):

TABLE 3.4.1B

Ethanol (land-using)	32 g
Biodiesel (land-using)	21 g
Ethanol (non-land using)	9 g
Biodiesel (non-land using)	9 g

....

FIGURE 3.4.1A. The greenhouse gas saving potential of different biofuels.

Note: this is based on 60 "well-to-wheel" life-cycle emissions studies, and shows a large range for each biofuel. It excludes land use change emissions.

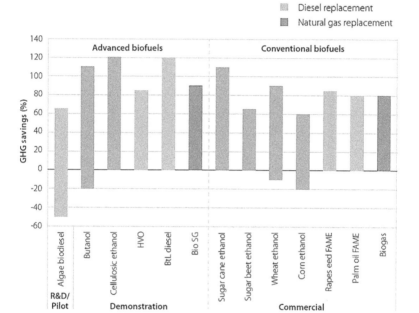

SOURCE: Bioenergy review (Committee on Climate Change, December 2011).

....

The IEA technology roadmap, *Biofuels for Transport*, provides an excellent overview of many of the fuelstocks and conversion technologies, from those in full commercial production now, to those which are just in R&D. It also examines their carbon balance and overall sustainability. It argues firstly that most conventional biofuel technologies need to improve

FIGURE 3.4.1B. Emissions from biofuels from various feedstocks.

Default value (CO$_2$/mj) of various fuels according to EU data

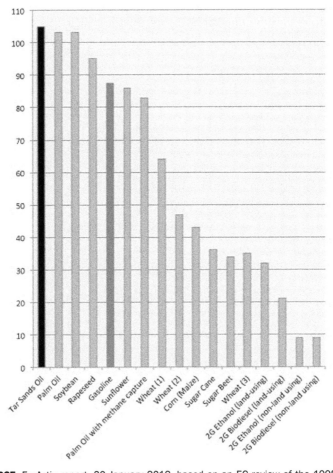

SOURCE: EurActiv report, 30 January 2012, based on an EC review of the 1998 Fuel Quality Directive.

conversion efficiency, cost and overall sustainability. Second, second-generation biofuels need to prove they can be commercial without their lifecycle impacts compromising food security and biodiversity, at the same time yielding positive social impacts. This includes sustainable land-use management and certification schemes, as well as supporting measures that promote 'low-risk' feedstocks and efficient processing technologies.

3.4.2 Indirect land-use change emissions

The unwanted emissions resulting from the change of use of land to growing biofuels are called indirect land-use change (ILUC) emissions. Growing palm oil or other fuels on former tropical rainforest or grassland can result in emissions hundreds of times worse than those from using fossil fuels. By contrast, growing dedicated energy crops on degraded land can transfer carbon from the air to the soil.

The industry is currently waiting for robust monitoring, standards and frameworks to account for indirect emissions. The European Union is attempting to address direct land-use emissions through employing sustainability criteria under the Renewable Energy Directive. These require that biofuels and bioliquids should deliver emissions savings of at least 35% over their lifetime compared to transport fossil fuels, rising to 50% in 2017 and 60% in 2018. This effectively rules out peatland, wetland and rainforest. Further criteria are expected, as well as standards and monitoring to ensure compliance.

The EU position on first-generation biofuels is being watched closely, since Germany, France and Italy are the EU's biggest producers of rapeseed oil, as well as possessing car manufacturing industries that actively support the use of biodiesel for reducing carbon emissions.

FIGURE 3.4.2. Indirect land-use emissions for different fuels.

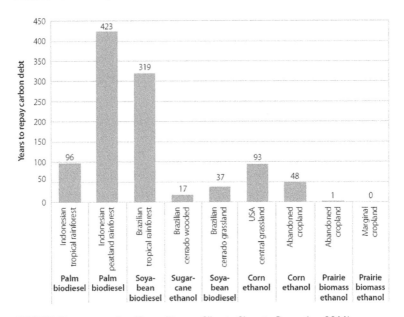

SOURCE: Bioenergy review (Committee on Climate Change, December 2011).

The Commission on Climate Change is advising the UK government to implement either the use of crop-specific ILUC factors, for example by adding estimates of emissions by crop, or through the setting of caps on the use of feedstocks with suspected unsustainable side-effects. Expect positive incentives to be offered for growing feedstocks with low ILUC risks. Potential investors should make sure they keep up-to-date with the latest developments in this field.

3.4.3 Existing sustainability schemes

Task 40 of the IEA Bioenergy Implementing Agreement has recorded the existence of 67 local, regional or global initiatives to develop sustainability criteria and standards for biofuels. The most significant are:

- The Global Bioenergy Partnership (GBEP), an intergovernmental initiative from 23 countries and 12 international organisations that has developed a methodology for assessing the greenhouse gas emissions associated with bioenergy.

- The Roundtable on Sustainable Biofuels (RSB), a voluntary international group of farmers, companies, non-governmental organisations (NGOs), experts, governments and inter-governmental agencies that has developed a third-party certification system for biofuel sustainability criteria.

- The International Organization for Standardization, which is developing an international standard via a new ISO project committee (ISO/PC 248, Sustainability Criteria for Bioenergy).

- The International Sustainability and Carbon Certification System (ISCC), which has developed an internationally recognised certification system for all kinds of biomass, including feedstocks for bioenergy and biofuel production.

3.4.4 How much land would we need?

An IEA report calculates that the world would need around 100 million hectares (1 million square kilometres), or over four times the area of the United Kingdom, to grow all the feedstock for the biofuels it expects we would need in 2050 to satisfy a demand of 145 exajoules (EJ – 10^{18} joules). It would come from crop residues and wastes, along with sustainably grown energy crops. This is a considerable challenge given the competition for land with food and fibre. The IEA concludes it should be possible, but that it may well require a change of diet on behalf of much of the global population so that less meat is required, since animal feed is responsible for much grain and soya bean consumption.

3.4.5 The water challenge

This transition to vegetarianism is likely to happen in any case, scientists have warned, to avoid catastrophic shortages of water. A 2012 report from the Stockholm International Water Institute (SIWI) says that there will not be enough water available on current croplands to produce sufficient food, let alone energy, for the expected nine billion global population in 2050 if current trends continue.

3.4.6 Social justice and biomass cultivation

A further constriction on biofuels expansion is the social dimension. The authors of *Biofuels, Land Grabbing and Food Security in Africa* (ZED Books, 2011) call for social as well as environmental aspects of sustainability to be taken into account by potential investors who may be tempted by offers that involve the purchase of land in developing countries for the growth of biofuels or other resources. This has especially been a recent tendency in Africa, where new economic and production alliances have been created that exclude the local, rural population. Furthermore, author Rune Skarstein uses existing research on Brazil's biofuel sector to suggest that long-term CO_2 emissions from the production and use of biofuels are higher than that of fossil fuels.

3.4.7 Ten sustainable solutions

To address all of these issues, several solutions are being developed:

1. Improving the use of wastes and residues as feedstocks to increase the efficiency of biomass use and avoid potential competition for land with agriculture or forestry.

2. Improving crop yields.

3. Introducing new energy crops, such as switchgrass and algae, and resistant crops that need less water, herbicides, pesticides and fertilisers.

4. Using dedicated energy crops to restore degraded or contaminated soils and store carbon in the soil.

5. Complying with the latest sustainability certification schemes.

6. Taking advantage of multi-season planting and intercropping, such as Integrated Food and Energy Systems, to minimise the amount of land and inputs needed to meet fuel, food and animal feed requirements and reduce the competition between them. This can also provide farmers with diverse income streams and promote the efficient use of residues and waste

7. Maximising conversion efficiency.

8. The use of renewable energy.

9. The deployment of biorefineries (see below).

10. Improving the energy efficiency of the facility.

TABLE 2. Land-use efficiency of different biofuel crops and expected yield improvements (global averages)

Biofuel type	Yields, 2010 (litres/ha)		Average improvement per year, 2010–2050	Resulting yields in 2050 (Lge or Lde/ha)	Main co-product, 2010 values (kg/Lbiofuel)
	Nominal	Lde/Ha			
Ethanol – conventional (average yield of feedstocks below)	3300	2300	0.7%	3000	
Sugar beet	4000	2800	0.7%	3700	Beet pulp (0.25)
Corn	2600	1800	0.7%	2400	DDGS (0.3)
Ethanol – cane	4900	3400	0.9%	4800	Bagasse (0.25)
Cellulosic-ethanol – SRC	3100	2200	1.3%	3700	Lignin (0.4)
Biodiesel – conventional (average yield of feedstocks below)	2000	1800	1.0%	2600	FAME: Glycerine (0.1)
Rapeseed	1700	1500	0.9%	2,100	Presscake (0.6)
Soy	700	600	1.0%	900	Soy bean meal (0.8)
Palm	3600	3200	1.0%	4800	Empty fruit bunches (0.25)

SOURCE: *Technology Roadmap: Biofuels for Transport* (IEA, 2011).

Case study: BP and sugarcane

BP has announced it is to invest heavily in expanding sugarcane cultivation and processing to produce bioethanol in Brazil and possibly southern American states. The decision is based on its 2030 Outlook report, which shows that the amount of crude oil the world needs to keep running will rise from 85 million barrels per day now to between 102 and 105 million barrels by 2013. Yet, oil finds are decreasing, especially in non-OPEC countries.

It projects that the proportion of biofuels in all transport fuel must rise from today's 5% by volume to 10% by 2030. Growing crops to provide this will require a great deal of acreage. BP has already invested several billion pounds in this field to date and plans significant further investment, but recognises that to be successful it needs materiality and scalability, i.e., large continuous tracts of land. Cost-competitiveness suggests the most attractive option is therefore Brazilian cane ethanol, because of the cheap availability of land and labour, and its 40-year history of producing the fuel. Sugarcane-derived bioethanol currently sells at $60 per barrel compared to between $80 and $100 per barrel for oil, without subsidies.

BP is acquiring first-generation conversion processing facilities in Brazil, but sees the future in lignin cellulosic conversion to ethanol and hopes to announce its first conversion facility in 2013, which will be the largest in the world. Tens of thousands of acres of land will be required and the entire plant will be converted to fuel.

BP is following Shell who, in 2010, established Raizen, a joint venture with the Brazilian ethanol company Cosan. In a landmark agreement for sustainability made in August 2012, Raizen promised not to source sugarcane from farmers who have encroached on the lands of the Guarani people in Mato Grosso do Sul state. Raizen produces 2.2 billion litres of ethanol per year mainly as a fuel for cars. It says it wants to use its withdrawal 'as a good example for other companies to follow'.

Whether BP will do so remains to be seen. BP claims that sustainability is at the heart of its plans, but in doing so speaks of fair treatment for its workforce, and the provision of rural development and rural jobs, not land rights and biodiversity.

Since ethanol does not have the same energy density as gasoline, further ahead BP is looking at creating biobutanol, which looks like gasoline and can be blended 60% with gasoline. It is working with DuPont to develop production processes. A facility in Hull in northeast England is already producing some, and some of it powered the BMW fleet during the London Olympics 2012.

3.5 Promising feedstocks

3.5.1 Cellulosic ethanol

This second-generation fuel is derived from lignocellulose, a structural material that comprises much of the mass of a plant. Feedstocks include switchgrass, corn cobs, woody and agricultural biomass. As BP has found, the cost of processing the raw materials still need to fall. A process called enzymatic hydrolysis is being commercialised at new facilities like GraalBio's plant in Brazil. This uses the latest enzymes BP has licensed from Novozymes and DSM. The race is on in labs around the world to produce the most efficient enzymes, patent them and license them out. However, many parts of the process need to improve, including harvesting and/or collecting biomass, which adds an average of $15 per tonne in costs.

3.5.2 Jatropha

FIGURE 3.5.2. Flowers of a jatropha plant.

SOURCE: Creative Commons.

Jatropha is a genus of plant whose seeds contain 35% oil that can be used for making biodiesel and jet fuel. It has been tested by Japan Airlines, Air New Zealand, Continental, Brazil's TAM Airlines and Mexican carrier Interjet with Airbus. A peer-reviewed study from Yale University has calculated that it could reduce greenhouse gas emissions from flying by up to 60%. 140,000 formerly impoverished farmers in India are now cultivating the crop on marginal land without compromising food supply or food pricing. It also meets the European Union's sustainability guidelines for second-generation biofuels.

However, a study for the NGOs ActionAid and the RSPB of a proposed 50,000-hectare jatropha plantation in Kenya found that emissions from producing the biofuel would be 2.5 to six times higher than the fossil fuel equivalents. The research examined the production lifecycle: land clearance, planting, harvesting, refining and transportation to Europe, all issues investors need to consider. Jatropha (or similar plants) can be grown on marginal land, but yields are lower, which may not satisfy foreign investors, but may be sufficient for local needs. Why not supplant opium poppy fields in Afghanistan with such plants?

3.5.3 Algae

Algae production offers many potential advantages. These micro-organisms produce biodiesel by absorbing carbon dioxide, thereby helping to combat climate change, while, depending on the process, processing sewage or absorbing greenhouse gases emitted by industrial plants. Co-products include fertiliser and clean water. Moreover, many algal strains have high oil content compared to other feedstocks, they consume little water, and can be cultivated on non-arable land. The route from the original oil to producing biofuel is well understood and similar to the production of vegetable oil.

The major challenges in bringing them to commercial status relate to the cultivation and harvesting of the product, and the extraction of the oil at a competitive price. Some studies made of the lifecycle emissions of micro-algal cultivation have shown that in practice they may not yield significant greenhouse gas emission savings because of the energy inputs required or the high mineral fertiliser use required, but these issues could be addressed.

A further concern identified by the UK Committee on Climate Change is that atmospheric carbon dioxide cannot be absorbed into intensive algal mass cultures fast enough to enable high growth. The sources of the gas must therefore be the combustion of fossil fuels or biomass during power generation or industrial processes, thus representing a form of carbon capture and storage.

A summer 2012 report from Lux Research concludes that, at the current stage of development, algae cultivation suffers from high capital costs at an industrial scale: an average of $202,000 per hectare, which yields a 48% loss. Algal biofuels therefore only stand a chance in niche markets, for example aviation, where the availability of low carbon alternatives is limited, or by producing high volumes of comparatively low-value biofuel alongside high-value co-products.

3.6 Prospects for biofuels

The IEA roadmap says that scale and efficiency improvements will reduce costs to make most biofuels competitive with fossil fuels by 2030.

Improved harvesting technologies can cut feedstock costs by $25 per tonne, according to Lux Research's report 'Pruning the Cost of Bio-Based Materials and Chemicals'.

3.6.1 Biorefineries

This report foresees the development of biorefineries, which, in the same manner as conventional oil refineries, produce a variety of fuels and other products from certain feedstocks, in order to maximize opportunities for returns. They may be divided into two categories: those which concentrate on producing biofuels, and those which concentrate on producing other materials with biofuels as a co-product. Biorefineries can use a variety of feedstocks. Their development is supported by the IEA Bioenergy Task 42.

Production costs from 2010 to 2050 in this roadmap seem high: $11 trillion to $13 trillion. But the marginal savings or additional costs compared to using gasoline/diesel are only +/-1% of total costs for all transport fuels, and this is without factoring in the externalities of environmental damage avoided from not using fossil fuels.

3.6.2 Micro-organisms

As noted above, research is progressing fast on the discovery and development of new microbes and production environments that produce biofuels more efficiently. For example, in July 2012 alone, researchers at Michigan State University have discovered bacteria which increase by 10 times to 35% efficiency the amount of energy that can be recovered from the fermentation of agricultural and other organic wastes to produce ethanol, while other researchers at Solarvest announced they have genetically engineered algae to continually produce hydrogen gas using light and carbon dioxide.

3.6.3 Bio-SNG

Bio-SNG fermentation has even better new product potential. Its many products, proven at lab scale or larger, include: ethanol, butanol, acetic acid, butyric acid, 2,3-butanediol and methane. Leading start-ups in this domain include ZeaChem, which is collaborating with Procter & Gamble, and LanzaTech.

A company called ECN is developing a system for the conversion of dry lignocellulosic biomass into natural gas quality bio-SNG, claiming 70% efficiency. Its demonstration 10 MW CHP plant will operate from 2012.

3.6.4 Road freight

Freight accounts for roughly 21% of emissions from the transport sector and 6% of total emissions from all sectors in the UK. Governments everywhere recognise that they should support the introduction of low carbon trucks and lorries in order to tackle this. The sector offers a number of opportunities for investors. Different countries have different strategies. The case study below illustrates the sheer variety of possible ways to use different biofuels in combination with other technologies.

Case study: The Low Carbon Truck Demonstration Trial programme

This UK initiative, announced in August 2012, unites fleet operators, vehicle convertors, gas hub providers and universities in a £23 million set of trials managed by the Technology Strategy Board, in partnership with the Department for Transport and the Office for Low Emission Vehicles. The real-life trials by 13 companies of over 300 low-carbon commercial vehicles should increase road haulage operators' confidence. The funding will help them run fleets of alternative and dual-fuel heavy-goods vehicles (HGVs) by meeting part of the difference in capital cost between traditional vehicles and their low carbon equivalents, as well as the cost of the refuelling points, including 11 new public-access stations around the country for use by any operator. Operators often cite the lack of refuelling infrastructure as a barrier to the take-up of alternatives to diesel.

The trials involve a variety of approaches:

- the John Lewis Partnership, together with its partners, aims to obtain a 70% cut in carbon emissions in a wide range of articulated vehicles by combining recent research into truck aerodynamics with technology that substitutes the majority of the diesel used with biomethane.

- G-Volution will trial 10 44-tonne dual-fuel commercial HGVs using their patented dual-fuel technology 'Optimiser' and biomethane. The articulated trucks, converted to dual-fuel, will be trialled alongside diesel equivalents, providing direct comparison data for different operating environments.

- United Biscuits will explore the value in powering 44-tonne articulated vehicles with used cooking oil.

- J.B. Wheaton and Sons Ltd will trial, with other fleet operators, the use of 28 vehicles fuelled from compressed natural gas or liquid natural gas, blended with renewable biomethane, to run dual-fuel gas converted trucks. Seven fixed refuelling stations and five mobile stations will be shared with other fleet operators.

- Robert Wiseman Dairies, collaborating with Chive Fuels, Cenex and MIRA, will test 40 dual-fuel 40-tonne articulated trucks substituting diesel with natural gas from two upgraded public-access liquefied natural gas stations.

The fleets will be run for two years. The data gathered will be analysed by the Department for Transport, after which results will be made available.

3.6.5 Aviation fuel

We have seen that both algae and jatropha are being investigated by aviation companies keen to find ways to reduce their greenhouse gas emissions. Trials are likely to continue, but because of the expense, biofuels will for some time only form a small part of total fuel requirements. The involvement of Virgin Atlantic is described in a case study below. Some companies are also looking at the use of refined, reclaimed cooking oil.

Amongst these are UK-based Thomson Airways and KLM Royal Dutch airlines, which have both launched flights that use aviation fuel derived from cooking oil, and, in China, the Zhejiang-based Sinopec Zhenhai Refining & Chemical Co. Their plant has a production capacity of approximately 20,000 tonnes per year, and is expected to be ready for reviews by China's aviation authorities in January 2013. However, the cost of refining waste cooking oil is still 1.5–2 times higher than that of regular jet fuel, and an annual production capacity of 20,000 tons is a tiny proportion of total demand.

Biofuel-powered airlines to and from Europe will be exempted or partly exempted from the European Union's carbon tax for airlines, a practice that may encourage carriers to use more biofuel in their flights, but only if the use of biofuel is cheaper than paying the tax.

3.7 Partnerships, mergers and acquisitions

The last 10 years have seen hundreds of start-ups, corporations, financiers and universities attempting to develop second- and third-generation technologies. As these approach commercial scale, for these start-ups, alliances with appropriate partners are vital. The emerging partnership networks are increasingly dominated by the same giant multinationals that control the traditional petroleum industry. These fuel producers, refiners and distributors are playing both sides, being both the competition and holding the keys to the mass market.

Significant companies leading the way to market are Gevo, LanzaTech, Amyris, and Solazyme, for the simple reason that they have been able to forge robust partnerships with these oil and gas majors, such as Valero, Shell, BP and Chevron, as well as having developed cost-effective, proven technologies. In total, in this field there are around 800 partnerships among 753 companies, where almost every relevant company is connected to another. Seventy-nine percent of these companies are connected in one large web. Of these, biological technologies form the most prominent nodes. Many technologies are represented, but the few biological processing companies, such as LanzaTech, Amyris, and Gevo, have built the more extensive partnership portfolios. Technology developers, such as Novozymes and Amyris, and next-generation ethanol players, like Mascoma, Coskata, and LanzaTech, form their own mini-networks. These own patents for biological processes that are inherently flexible; organisms capable of consuming several feedstocks and producing several end products.

3.8 Structural challenges

Several structural challenges need to be addressed for biofuels to play their expected part:

1. The dismantling of trade barriers is an important prerequisite for the development of international trade between biomass-rich regions and biofuel production/consumption centres. Over the next few years, tariffs must be reduced and eventually abolished, argues the IEA.

2. The international technical standards for biomass, biofuels and intermediate products (e.g. pyrolysis oil) need to be harmonised internationally.

3. Robust sustainability certification for biofuels is needed, that is not biased towards certain regions or technologies, especially for developing countries.

4. International trade agreements should be made to stimulate the production of biofuels for export, especially in developing countries.

Case study: LanzaTech and Virgin Atlantic

A glimpse into this world is provided by looking more closely at LanzaTech, the chief winner of funding from the US Federal Aviation Administration (FAA), which is supporting the development of alternative and sustainable sources for commercial jet biofuel. In December 2011 it awarded contracts to eight companies for R&D, and LanzaTech received the lion's share of $3 million.

LanzaTech was named in March 2012 as one of 10 New Energy Pioneers at the fifth annual Bloomberg New Energy Finance Summit in New York, because of their 'innovative, proven technologies, robust business models and the ability to demonstrate traction and global scale'. Virgin Atlantic has signed a deal with it to produce kerosene jet fuel from ethanol using its technology, to power flights from Shanghai and Delhi to London. The first commercial production facility is anticipated in China by the end of 2013. LanzaTech is 51% owned by Khosla Ventures of Silicon Valley in California, and was founded by Vinod Khosla, the co-founder of Sun Microsystems.

Its fermentation technology uses a patented microbe that grows on syngas or methane offgases from many industrial processes, and converts them into ethanol. LanzaTech is also working with French company Global Bioenergies to produce isobutylene from waste carbon monoxide, which currently has to be expensively scrubbed from many industrial processes. If successful, this will be significant, because nowhere in nature is isobutylene made using bacteria. Isobutylene is used to make fuel additives and kerosene: jet fuel.

LanzaTech is installing its technology on a steel mill in South Auckland, which will be operational next year. Potentially, it could be implemented relatively quickly on a wide scale throughout several industries. It also has a deal with Mumbai-based renewable energy investment company Concord Enviro Systems.

The process is not unique. Several companies throughout the world are also competing, whether small innovators with large partners: BP and Verenium, OPX Bio and Dow; or two large companies together: BP and Dupont, and Rentech and ClearFuels. Californian company Fulcrum Bioenergy has patented processes to turn municipal solid waste into syngas using gasification, then catalytically convert it into bioethanol. Its first plant, east of Reno, Nevada, should be up and running in 2013, having received a $105 million conditional US Department of Agriculture loan guarantee. It could produce up to 10 million gallons per year.

CHAPTER 4

Electric Vehicles

ELECTRIC VEHICLES (EVs) – smooth, silent, easy to drive – range in type from those running purely on electricity from their batteries, at one extreme, through plug-in hybrids (PHEVs), to electric/ICE hybrids. At the lowest level of the continuum is the simple stop/start level of electrification (dubbed micro-hybrids), designed to make between 10 and 15% of fuel savings by cutting fuel wastage at traffic lights and in traffic jams through the use of technologies like Ford's Smart Regenerative Charging.

4.1 **Hybrids**

Hybrids are a fairly common sight on the road now. Ford, for example, has more than 200,000 in operation. Their chief advantages are using battery power to let the combustion engine run at its optimum efficiency, and reclaiming energy to deliver greater fuel economy. They first appeared in the United States, and as they become cheaper, we'll see more and more of them in Europe.

Hybrid vehicles represent a route towards pure electric vehicles, but PHEVs have the advantage of longer range. Here, the internal combustion engine works as a back-up when the batteries are depleted, which gives confidence to drivers. With the present, just emerging, networks of electric vehicle charging points, electric vehicle range extenders alone represent a poor safety net for those fearing being stranded without charge.

Ayoul Grouvel, head of Peugeot electric vehicles projects, believes by 2030 the majority of cars will be electric, with some form of hybridization being dominant. This view is shared by Peter Mock, an environmental scientist at the International Council for Clean Transportation.

4.2 **The advantages of EVs**

Moving on to pure EVs, their chief selling point for consumers is fuel economy: whereas diesel cars cost an average of 13 p per mile, due to the price of electricity, electric cars work out at an average of 3 p/ mile. Other benefits include: exemption from annual vehicle tax, discounted insurance premiums, reduced company car tax, parking concessions offered in some areas, and, in London, 100% discount on the congestion charge. For local administrations, further advantages are the lack of noise and on-street emissions, which helps bring down levels of air pollution.

Proponents of electric vehicles argue that 'the future is electric', making them more compatible with modern lifestyles; the public increasingly expects vehicles to contain features such as IGS and navigation, content streaming, mobile hotspots, application downloads and wireless charging. However, the future widespread adoption of EVs is dependent upon sufficient installation of new and additional low carbon generation infrastructure to supply the increased demand. There will be plenty of space in the market for vehicles powered with other fuels for some time to come.

Presently, the maximum distance driveable on a single charge is around 150 km. An OECD study found that many urban households may therefore purchase an electric vehicle just for local use. This may be a two-wheeler or other small, purpose-built, low range, agile and congestion-beating vehicle. An additional larger vehicle might be owned for long or family journeys.

The further an EV travels each day, the more cost-effective they are to their owners, as long as they are frequently charged. Another study by the International Transport Forum found that electric passenger cars currently cost €4000–5000 more to their owners than an equivalent fossil fuel car over the vehicle's lifetime, but because they will travel greater distances, an electric delivery van costs €4000 less to owners over its lifetime than a similar fossil-fuel van. However, the costs to society of electric cars and vans today is €7000–12,000 more than fossil-fuelled equivalents; this figure will come down over time as the infrastructure is installed.

Recent estimates put the global electric vehicle fleet at over 120 million units (mostly electric bicycles and scooters in China) with the most popular electric car models achieving global sales of approximately 44,000 units in 2011. Electric bicycles and scooters are ideal for single commuter traffic in urban situations, reducing congestion as well as air pollution. Ford Europe expects EV sales to form up to a quarter of all sales by 2020. The biggest share will be hybrids and plug-in hybrids. Pure EVs will be no higher than 5%, they estimate.

4.3 Carbon impacts

In almost all cases, EVs will generate fewer lifecycle CO_2 emissions than comparable ICE counterparts (Source: Electric Vehicles Revisited: Costs, Subsidies And Prospects, Discussion Paper No. 2012-03, Philippe Crist, International Transport Forum, Paris, France, April 2012). Exactly how much less depends on the carbon intensity of marginal electricity production used to charge them, the electricity generation mix in the country concerned, the full lifecycle emissions (including production) of comparable electric and fossil-fuel powered vehicles (and their fuels), and the relative energy efficiencies of those vehicles. In most scenarios, the marginal CO_2 abatement costs of replacing fossil-fuel powered cars with EVs remain high, with the exception of high mileage vehicles.

This discussion paper, 'Electric Cars: Ready for Prime Time?', produced for the OECD, compared the lifetime impacts of different vehicles, taking into account various electricity generation scenarios. It found that for four-door sedan and five-door compact cars the cost of using EVs to combat climate change was at the high end of the range of costs of all measures to reduce carbon dioxide emissions in the transport sector: between €500 and €700 per tonne of CO_2 avoided. That is a lot. As with running costs, the compact electric van was more of a bargain because it travels further. If the local electric grid contains a high proportion of coal-generated power, then there may be no carbon dioxide savings over conventional vehicles. The study says: 'even in regions where baseload generation is relatively low carbon, high rates of peak hour charging will come from marginal electricity generation which may be much more carbon intensive', like coal, oil or gas. Therefore, 'the timing of recharging will have a significant impact on overall greenhouse gas emissions for electric vehicle use'.

FIGURE 4.3. Relative journey lengths and carbon emissions for 22 electric Mitsubishi i-MiEVS and conventional cars.

Relative journey lengths and carbon emissions for 22 electric Mitsubishi i-MiEVS and conventional cars, from a real-world trial run in 2010 by the Technology Strategy Board's £25 million Low Carbon Vehicle Demonstrator trial at Aston University, Birmingham, UK. The data show that an average overall distance of 23 miles was travelled each day by drivers, giving 'a reassuring margin for flexibility should longer journeys be required' according to Neil Butcher, Arup's project leader of CABLED – Coventry and Birmingham Low Emission Demonstrators – a consortium made up of 13 organisations, led by Arup. The vehicles were left parked for around 97% of the time, with usage particularly low during school hours and overnight.

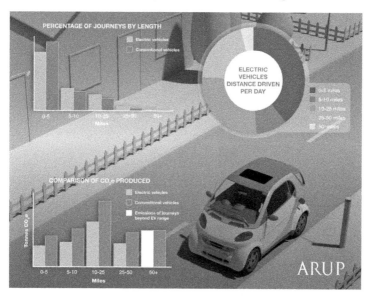

SOURCE: ARUP.

4.4 Vehicle charging

We'll come back to the issue of the optimum times of day to charge EVs. First, let's look at the four basic ways of charging an electric vehicle:

TABLE 4.4. Charging point types/locations

Type	On-street/ off-street	Public/ private	Locations	Shared	Restricted access (RA)/Open access (OA)	Plug type
A	On-street	Public	All (high streets, residential areas, etc.)	Yes	RA	3-pin
B	Off-street	Public car parks, leisure & retail centres, community facilities, stations, parks	Yes	RA/OA	3-pin/blue commando	
C	Off-street	Private car parks, workplaces, residential apartments	Yes	RA/OA	3-pin/blue commando	
D	Off-street	Private	Residential	No	OA	3-pin/ blue com- mando

Note: Restricted access (RA) means the use of the points is restricted to subscribers.

Mode 1. Domestic AC socket with extension cable

Here, the electrical installation must comply with safety regulations and have a residual current device (RCD) and earth leakage protection. This is the simplest solution. It uses three-pin domestic-style square plugs/sockets certified against BS1363 (rated 13 amp). (This mode is not permitted in some countries.)

Mode 2. Domestic AC socket, dedicated cable with protection

A dedicated socket is provided on a single-phase or three-phase network, with the same safety requirements. This solution is more expensive. This and the following two modes can use a three-pin plug or industrial-style blue commando five-pin plugs/sockets (rated 16A, 32A or 63A), certified against the international standard IEC60309. Charging points equipped with Type 2 (e.g. EN 62196-2) seven-pin sockets can only be installed outside at domestic properties.

Mode 3. AC charging point

As for mode 2, but with the charging control system function on the vehicle charging device and an EVSE control module in the charging installation.

Direct current (DC) connection

Similar to mode 3, but the AC is converted to DC, which the battery prefers, drastically speeding up the charging time. Requires an AC/DC-sensitive RCD on the network side.

4.4.1 Standards

Traditional AC electric vehicle charging points can be installed anywhere. The general requirements for both are similar: a tethered cable and plug (e.g. SAE J1772) installed inside or outside the property. Charging points are typically rated at 3.7 kW (16 amps) or 7 kW (32 amps) power output and can reduce charging times by about two and four hours respectively, compared to a standard 3 kW household three-pin BS1363 connection.

4.4.2 How to charge an EV

FIGURE 4.4.2A. A typical wall-mounted domestic or workplace charging point with a tethered Type 1 five-pin (SAE J1772) connector.

FIGURE 4.4.2B. A typical workplace (or external domestic) charging point with a Type 2 (EN62196) seven-pin socket outlet.

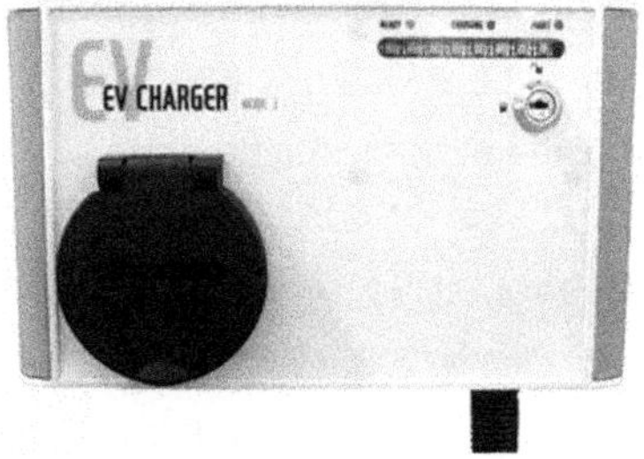

SOURCE: Cenex.

SOURCE: Cenex.

VIDEO: HOW TO USE A QUICK CHARGER

http://www.youtube.com/v/u0-irj5idmi&hl=en_ us&feature=player_embedded&version=3

4.4.3 Which type of charge point should you have?

Using dedicated fast or quick/rapid charging equipment offers distinct advantages over using a standard three-pin UK household connection (BS1363):

1. They can handle higher electrical currents for longer and are IP rated to withstand weather.

2. A 32 amp (7 kW) single phase AC fast charging point can charge an EV typically twice as fast. A 50 kW DC fast charging point can charge even faster, achieving 80% State Of Charge (SOC) in 30 minutes.

3. Fast charging points may contain additional features, including a timer and energy meter to monitor electricity use and take advantage of lower rate night-time or EV tariffs.

4. Fitting a dedicated charging point with a high capacity connection will ensure future upgrades to higher power, faster chargers is made easier.

5. Charging points (depending upon their type and location) can be connected to a Charge Point Management System that provides charging data capture, allowing users and charging point hosts to monitor charging, and can display the location of other nearby charging points.

6. Many charging point and energy suppliers offer products that include installation and care packages, taking the stress out of owning a charging point.

Dedicated fast charging points typically resemble an on-street bollard (when ground-mounted) or a wall box similar to a hose reel (when wall-mounted) and provide either a tethered Type 1 (SAE J1772) five-pin plug or a socket(s) capable of accepting Type 2 (EN62196-2 compliant) seven-pin plugs. Fast charging points usually provide 7 kW power at 32 amps.

Direct current fast charging points are suitable for motorway services and private or public car parks. Dedicated quick/rapid charging points typically resemble a forecourt petrol dispenser and require three-phase electricity. The unit comprises high power electronics in a weatherproof housing with a heavy duty tethered cable and plug (JEVS G105-1993 or JARI Level 3, CHAdeMO Association compliant). Quick/rapid charging points provide 50–70 kW DC power at about 125 amps. They offer a convenient and rapid means of charging a battery from flat to 80% in up to 30 minutes.

They are compatible with several vehicles including the Citroen C-Zero, Mitsubishi i-Miev, Nissan Leaf and Peugeot iON (more compatible vehicles will be available soon). The equipment communicates with the vehicle onboard systems to establish whether it is safe to charge before power is transferred. The connector is physically locked into the vehicle socket. They need planning permission to install if over 1.6 m in height.

4.4.4 Battery charging times

TABLE 4.4.4. Battery charging times dependent on charging method

Power supply	Voltage	Max. current (amps)	Charging time (hours)
Single phase 3.3 kW	230 AC	16	6–8
Single phase 7 kW	230 AC	32	3–4

Three phase 10 kW	400 AC	16	2–3
Three phase 24 kW	400 AC	32	1–2
Three phase 43 kW	400 AC	63 A	0.3–0.5
Direct current 50 kW	400–500 DC	100–125	0.3–0.5

4.4.5 Induction charging

A future alternative that will remove the need for leads is wireless EV inductive charging (WEVC). This merely requires the vehicle to be close

FIGURE 4.4.5A. Inductive charging in a parking space.

SOURCE: Qualcomm.

FIGURE 4.4.5B. The principle of inductive charging.

① Power Supply ④ Receiver Pad

② Transmitter Pad ⑤ System Controller

③ Wireless Electricity & Data Transfer ⑥ Battery

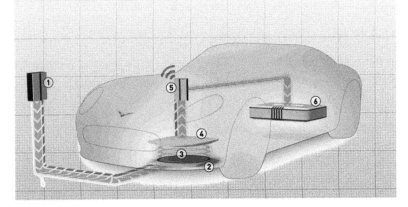

SOURCE: Qualcomm.

to the charging point rather than physically connected (similar to an electric toothbrush). Options include stationary charging and mobile charging. Mobile charging, where the inductive cables are buried beneath the road surface and vehicles draw charge as they travel, has been tested in trials. Stationary systems are located beneath a parking space. One technical advantage is that it avoids the need to deep discharge the battery, keeping it more or less topped up. This would extend battery life.

Various companies are seeking solutions that do not need accurate alignment of the vehicle over the charge point, and can deliver high energy transfer with little power loss over a wide area gap. Amongst these is Qualcomm, with the Halo WEVC technology it acquired from buying up New Zealand-based HaloIPT in November 2011. Various power transfer levels have been proven in trials from 3.3 kW to 20 kW systems for a racing car. Power transfer efficiency is comparable to plug-in cable charging systems; there are losses in both. A key advantage of below-the-surface inductive charging is that it makes the process of charging as simple and convenient as possible for drivers, which will help to push the introduction of electric vehicles. Dynamic charging means longer distance journeys and smaller batteries and lighter cars. A universal open standard, compliance and a regulatory environment are yet to be achieved; these would propel its introduction and ensure interoperability.

In London, Qualcomm envisages a two-year phased approach to the introduction of inductive charging. The first phase will see deployment for fleets and privately owned vehicles, with the second, broader, phase taking in taxis and taxi ranks and commercial fleets.

4.5 Battery development

Much work is ongoing to improve battery storage. Most experts agree that prices for electricity storage will fall in coming years, but disagree over how far and how quickly. A significant drop in prices could have wide-ranging effects across industries and society itself. Cheaper batteries would enable the broader adoption of electrified vehicles, potentially disrupting the transportation, power and petroleum sectors.

Research published in July 2012 by McKinsey indicates that the price of a complete automotive lithium-ion battery pack could fall from $500–

600 per kilowatt-hour (kWh) today to about $200 per kWh by 2020 and to about $160 per kWh by 2025. It is not in the auto industry that the first advances will be made, but in the consumer electronics industry, where competition is even more intense to improve battery efficiency. In the United States, where gasoline prices are at or above $3.50 a gallon, automakers that could acquire batteries at prices below $250 per kWh would be able to offer electrified vehicles at a price equivalent to vehicles powered by advanced internal combustion engines, on a total-cost-of-ownership basis. In Britain, where fuel costs more, this would be even more attractive.

Two technical developments in batteries are required for the widespread adoption of electric vehicles: charging times must be reduced significantly, and storage capacity increased dramatically. There have been some exciting developments in this respect: at the beginning of last year, the University of Illinois and nearby Northwestern University announced a breakthrough in charging time, and last October, Nissan, working with Kansai University in Japan, announced that it had reached a charging time of just 10 minutes. These and similar solutions are lithium-ion batteries and use changes to the design of electrodes. Nissan says it will take up to a decade to get them to the marketplace. One of the main challenges is to minimise the reduction in capacity of the battery over time that's caused by fast charging and frequent discharging.

There are several competing paths to develop cheaper batteries; one is to reduce the amount of expensive and comparatively rare metals required in components. A second is to develop completely new battery designs that don't need such metals. Manufacturers are developing high-capacity anodes made of silicon, a common material, that would be 30% more efficient than today's graphite anodes. A chief strategy here is to incorporate multilayered cellular structures within batteries that have the effect of eliminating 'dead zones'. Researchers are also developing cathode-electrolyte pairs that they hope by 2025 will be able to increase cell voltage to 4.2 volts from the present 3.6 volts, an increase of 17%. Such technical advances in cathodes, anodes and electrolytes could increase the total capacity of batteries by up to 110% within 7 to 12 years; this could account for almost half of price reductions.

4.6 Charging point roll-out

Central to mass take-up of electric vehicles will be the widespread availability of charging points. According to the website Open Charge Map, the UK ranks third in the world for the number of public charging locations. At the time of writing, 705 are registered. The United States tops the list with 4229, and Norway comes second with 952. The Netherlands has 697 and Germany 505.

A small industry is growing up providing charging points and associated support systems. Charge point solutions are available to plot owners who wish to install them with a variety of payment and support options including free-to-use, payment by coin, token, mobile phone, loyalty card, smartcard or credit card, and computerised or web-based management systems.

For those considering whether to install systems, whether local councils, employers or retailers, here is an approximate indication of the relative cost level for different charging point types and locations:

TABLE 4.6A. Costs for EVCP equipment and installation

Type	On-street/ off-street	Public/ private	Shared	Restricted access (RA)/ Open access (OA)	Plug type	Cost level	
						Equipment	Installation
A	On-street	Public	Yes	RA	3-pin	High	High
B	Off-street	Public	Yes	RA/OA	3-pin/ blue commando	Medium/ high	Medium/ high
C	Off-street	Private	Yes	RA/OA	3-pin/ blue commando	Medium/ high	Medium/ high
D	Off-street	Private	No	OA	3-pin/ blue commando	Medium/ low	Medium/ low

TABLE 4.6B. Cost indications for Table 4.6a

Cost level	Cost (approx.)	
	Equipment	Installation
Low	£0–500	£0–500
Medium	£500–1000	£500–1500
High	£3500–4000	£3000–5500

The range is due to the optional variables, with the most significant being whether the charging points are restricted access or open access types.

There are already a number of schemes, where, for a small subscription, drivers obtain access to a network of charging points, a range of services such as a charging cable, free parking in certain areas, and pay for the electricity they use. Presently, comprehensive charging coverage is only available in London, which boasts more charge points than conventional filling stations. But more and more charging points are appearing, partly as a result of the British government's Plugged-in Places programme.

4.6.1 Plugged-in Places

The UK government is supporting the roll-out of a network of charging points through its 'Plugged-in Places' programme by offering match-funding to consortia of businesses and public sector partners for the cost of the installation of electric vehicle recharging infrastructure. Delivered by the Office of Low Emission Vehicles, the plan is to have more than 9000 charge points throughout the UK, with the exception of Wales, by March 2013, with the first wave coming in London, the North-East and Milton Keynes. An additional 4000 charge points will then be rolled out to Scotland and Northern Ireland, Manchester, the Midlands and the east of England in the second wave. Wales is relying on the private sector to provide points. Charging points are also being installed by councils across the UK and by private sector providers.

Case study: Chargemaster

Chargemaster has jumped in to secure market advantage, seizing the incentive offered by the Plugged-in Places scheme of 50% match-funding for their own investment. It's setting up partnerships with local authorities who need a private sector buddy to do the business for them. Polar network, a division of Chargemaster, is hoping to set up 40 charging bays in 100 cities by 2013. Their mission is a network of charging posts in every major town across the country, in supermarkets, car parks, shopping centres, railway stations and airports. In the summer of 2012 it installed the first fast charging unit in a supermarket, Abingdon's Waitrose supermarket. It also has a partner in Britannia Parking to provide charging bays across its car parks.

All this activity is creating investment opportunities. Good partnerships are key to early success for intrepid field entrants.

Case study: Ecotricity and Welcome Break

Another early entrant was renewable electricity supplier Ecotricity who, in summer 2011, partnered with motorway service chain Welcome Break to install free charging points along motorways. Drivers can top-up in just 20 minutes using rapid 32 A recharge points, or fully charge in two hours; while those using the slower 13 A supply will be able to recharge fully if staying overnight in adjoining hotels. As it is free, a driver travelling a year's typical 12,000 miles could save up to £2000 in petrol costs at today's prices, and around 2500 kg in CO_2 emissions. Eleven of the service stations have this facility; the remaining 16 will have it soon. Ecotricity says the chargers have been used about 300 times in the first year, which it claims represents around one in ten of all electric vehicles on the market during that period. Ecotricity recognises that this is not a huge number, but it intends to be 'ahead of the curve in terms of the switch to electric vehicles, as both a learning process for itself and the early adopters of electric cars'. Before this, Ecotricity had already installed a charging post connected to its wind turbine by the M4 motorway at Reading; this was the first charging post to be powered directly from a wind turbine.

Rod McKie, CEO of Welcome Break, says his company 'wants to be at the forefront' of the coming change in motoring habits: 'as hybrid and electric cars become part of everyday life, Welcome Break will have the facility to fast-charge these cars, giving electric car drivers the opportunity to travel the length and breadth of the UK'. Naturally, it hopes this means more people will use Welcome Break's traditional facilities.

FIGURE 4.6.1. The Nemesis, an electric sports car developed by Ecotricity, which beat the British land speed record with an average speed of 151mph in 2012.

SOURCE: Ecotricity.

Case study: British Gas

In the English Midlands, British Gas is offering charging equipment to households as part of a package that also takes advantage of the Plugged-in Places grant funding. The offer is dependent on simple requirements, and comprises installation of a single wall box charger and one year's membership to every Plugged-in Midlands public charging point. Those taking up the offer also receive a Navetas Home Energy Monitor Kit that lets them monitor the electricity used by the electric vehicle and the whole house, plus three years' Home Electrical Care service.

4.6.2 Charging technology market leaders

Companies in the EV-charger space globally include corporations like ABB, Schneider Electric, Siemens, Eaton, Nissan, Fuji Electric, Aker Wade and Tokyo Electric Power Co. This is the utility that developed the standard fast-charging technology known as CHAdeMO that's now in use in almost all DC fast-charging systems, despite the fact that it competes with the Society of Automotive Engineers' SAE J1772 fast-charging standard and a German system using an adapter from Mennekes Elektrotechnik, known as VDE-AR-E 2623-2-2.

ABB's Terra 51 fast DC 50 kW charger can give a vehicle 50% charge in 15 minutes and takes 30 minutes for a full charge. Having been deployed in Europe since early 2012, it has recently achieved Underwriters Laboratories (UL) certification to begin sales in the United States. Its system uses technology acquired in 2011 from Dutch start-up Epyon for an undisclosed sum. It also deploys software devised by Ventyx, an electricity grid IT firm it bought for about $1 billion in 2010, to manage grid demand and produce usage reports. The software can also serve as a networking platform for partners, such as ECOtality and its Blink charger network, to deploy different business models via a set of APIs. ABB also makes 120-volt and 240-volt chargers for garage or office car park use, and last March launched a 20 kW DC charger, the Terra Smart Connect, that can charge an EV battery in about an hour. One of its projects is to cover the whole of Estonia with 100 DC–AC fast charge points at filling stations, and 500 mode 2 chargers in office and government car parks. Led by the country's Ministry of Economic Affairs, Estonia wants to create the world's first complete, nationwide EV charging infrastructure.

4.6.3 Winning strategies

These large players are acquiring smaller start-ups that have developed new technologies they believe they can exploit. Likewise, they themselves work with auto-manufacturers to determine their precise requirements for models under development. Recent deals of this nature have included the raising of $84 million by Protean Electric to build a factory manufacturing electric drive technology in China. Investors included GSR Ventures, New Times Group, Oak Investment Partners and the city of Liyang, Jiangsu Province, China. And in July Polaris Industries and other investors contributed $13 million as the first tranche of a $45 million funding round by Brammo, a developer of electric vehicle technology and a manufacturer of two-wheeled EVs. Several industrial giants and start-ups alike are exploring the potential for vehicle-to-grid technologies that turn plug-in cars from a liability to an asset for utilities.

AeroVironment and ECOtality in the US have deployed fast charge points, along with start-up Coulomb Technologies, with the latter two companies also making networking and software that links cars, drivers, utilities and other third parties to manage charge monitoring, payment methods and other such functions. Tesla is planning a network of 'supercharging' stations, along with batteries that are swapped for charged ones, instead of waiting to be refilled. This is a model that Israeli company Better Place is trying to push into the market, and which has been very successful in Israel, so much so that it has attracted investment from GE, UBS and others. Crucial to their success has been the level of customer service, which is reportedly giving high satisfaction. In a highly competitive market like this, the way the hardware, in this case cars, and the software, which

SUSTAINABLE TRANSPORT FUELS BUSINESS BRIEFING

interacts between the car, the utility, and the charging supplier, fits into customers' lifestyles is critical to success.

French charging point manufacturer DBT has partnered with UK-based ChargePoint Services to distribute its 'Quick Charger' technology. DBT's Rapid Charge unit uses ChargePoint's management system to provide monitoring and data analysis for billing and cost purposes. DBT also has a partnership agreement with Nissan Europe and Gateshead College in the north-east of England to launch a business to improve the supply chain around the electric vehicle market. The business model will be different in Europe from the United States. In Europe, this is usually between the EV driver or charge point operator, and their utility. But in the United States, Californian state regulations that prohibit utilities from setting up EV charging businesses themselves will drive the industry in a different direction, opening the doors for different business models.

In the US, a $230 million Department of Energy-backed EV Project is putting fast-chargers up and down the California–Oregon–Washington I-5 corridor and in a 425-mile stretch of highway between Nashville, Knoxville and Chattanooga that's known as the Tennessee Triangle. Charging platforms also include Coulomb's ChargePoint and ECOtality's Blink, which all require drivers to sign up in advance and get a smartcard or a passcode to provide access. While these early entrants have sought in this manner to get their foot in the market, other business models, such as using credit cards or Bluetooth via mobile phones, are likely to appear. For example, GE and PayPal have formed a partnership to provide online and mobile payment, and GM has opened its OnStar system to apps developers to give charging location and status data to drivers via their Volt dashboard.

4.7 **EV fleets**

In the UK in 2011, 58% of all new cars and 68% of vans were bought by fleets, making this market a significant take-up for new low carbon vehicle technology. For this reason, the UK Department for Transport and Transport for London are sponsoring the Energy Saving Trust, in partnership with EDF Energy, to run a 'Plugged-in Fleets Initiative', which is designed to promote the uptake of electric vehicles by fleet managers.

Twenty organisations have signed up to this initiative, including Boots UK, London Fire Brigade, Network Rail, Surrey County Council, Southwark Council, Tristar, the University of Cumbria, Wm Morrisons PLC and York City Council. They receive guidance and a strategic plan for the introduction of electric vehicles into their fleets.

The project has three aims. Besides providing a tailored report for each participating organisation, outlining the benefits, it will offer wider practical advice for all business fleets, thereby enabling fleet decision-makers to purchase and use electric vehicles where they work best.

EDF Energy is providing advice on vehicle charging, and software is provided by Route Monkey that shows which existing routes have the most charging points before the organisation makes a purchasing decision. Once the vehicles are in operation, the software, called EVOS, maximises their use and helps fleet operators achieve the best possible reduction in fuel and emissions. The final results from this pilot scheme will be available in January 2013.

4.8 **Smart grids and EVs**

Without action, the widespread take-up of EVs will increase the demand for electricity beyond supply capacity. As well as building new low- or zero-carbon generation plants, and investing generally in energy efficiency, smart grids are the third essential requirement for the eventual mass public uptake of EVs. A study by the OECD's International Transport Forum, published in July 2012, *Smart Grids and Electric Vehicles: Made for Each Other?*, finds that smart grid technologies, if rolled out nationally, make an ideal partner for EVs, because they enable demand management and therefore a reduction in peak electricity supply requirements, meaning fewer power stations would have to be built.

Moreover, EVs will be able to use their batteries to store intermittent solar and wind power supplies at off-peak times and feed the power back into the grid when needed, should the vehicle not require it. 'Vehicles are parked on average 95% of the time, providing ample opportunity for the batteries to be used in this way', the report observes. EV owners would be paid for permitting this, reducing the overall lifetime cost of owning such a vehicle. The report therefore recommends changing the pattern of electricity pricing to encourage smart grid technology take-up and the use of EVs for storing power. Vehicles could even provide backup in the event of power cuts.

This scenario would take off after 2020, the date for the UK's target of having smart meters in every building in the country. Many companies are positioning themselves to provide this service, and a huge amount of private capital is being invested. The digital technology installed in the metering network will enable communication, and, where permitted, switching capability, to be two-way between the utility and the customer's premises. Consumption and pricing information will be available almost in real time. Utilities will therefore be able to dynamically manage the system as efficiently as possible, minimising costs and environmental impacts, while maximising system reliability.

It has been assumed that off-peak priced electricity would only be available overnight, but the report suggests letting these tariffs apply automatically anytime the vehicle is charged. This will provide opportunities for EV owners, business fleet managers and employers, to profit from reselling electricity. It could either be returned to the grid, or used locally; software could be designed to determine, at any moment, which is the most financially advantageous. EV ownership could grow to

Case study: Quimera

Quimera is a Barcelona-based company with an innovative business model. It developed the All Electric Gran Turismo (AEGT), the first prototype of its kind, that was launched at the Frankfurt Motor Show in 2011. This achieves top speeds of 300 km/h from three electric motors, supplied by UQM Technologies and runs at 700 bhp. It is open about the technologies inside its cars. It has a partnership with French multinational engineering consultancy Altran, with which it has established a centre of excellence in Barcelona.

Away from motorsports, Quimera has interests in urban environments and mobility, with 12 projects, each with a global budget of €12 million, and joint venture agreements with organisations ranging from delivery company DHL to the City of London. It is supplying electric motorcycles for the City of London Police and the Barcelona Police Department, as well as electric commercial vans for retail businesses and urban wind turbines. It has plans for expansion in the UK, India, US, Italy and Brazil, and is already established in China. Currently it is making profits of about €400,000. It expects to make a million soon and then to go public.

Case study: GM Ventures

General Motors has a subsidiary corporate venture fund, called GM Ventures, which is selectively investing in electric vehicle technology. Initially funded with an overall $100 million, a subsector of its focus includes 'automotive clean tech'. As defined by GM, this includes technologies related to the greening and electrification of a vehicle, for example with emission controls, batteries, motors, smart grid and vehicle energy efficiency technology.

It has made 12 investments to date, but has yet to make one in Europe. In its first six months it made four investments, including in Powermat, an Israel-based company that is developing wireless charging technology. While interested in Asian markets, its director in Europe, Jernefa Loncar, says that investment opportunities in markets such as China tend to be more private equity/execution raise, so the main focus for GM Ventures tends to be currently in Europe, North America and Israel. It intends to be a backer of early-stage technologies in the European automotive sector.

One area of investment has been in solar energy systems charging with the provider ViSole. At the same time as that deal, it has signed commercial agreements with the company for the installation of solar charging canopies at Chevrolet dealerships and GM facilities.

account for a substantial share of electricity consumption, perhaps over 20% in the long term. The greater the increase in consumption, the more inevitable these technologies become.

CHAPTER 5

Fuel Cells

LOOKING BEYOND BATTERY-POWERED EVS, we glimpse a scenario long dreamed of by advocates of a totally renewable energy-powered future: a quiet, pollution-free, hydrogen-powered, fuel cell-driven world. This future is already making its presence felt.

5.1 What is a fuel cell?

A fuel cell converts the chemical energy in a fuel into electricity by reacting it with oxygen or another oxidising agent. Hydrogen is the most common fuel, but hydrocarbons like natural gas and methanol may also be used. They differ from batteries in that they will never run out as long as fuel continues to be supplied. Fuel cells were invented by the Welsh physicist William Grove in 1839.

Like batteries, they consist of an anode, a cathode and an electrolyte. Traditionally, in a proton exchange membrane fuel cell, the electrodes are made of platinum, which acts as a catalyst because it is highly reactive with oxygen. Individual cells produce only small electrical potentials of around 0.7 V, and are connected together or stacked in series to increase the voltage. Their efficiency is between 40 and 60%, dependent on the load.

FIGURE 5.1. How a proton exchange membrane fuel cell works.

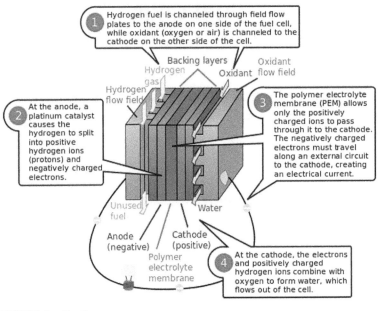

1. Hydrogen fuel is channeled through field flow plates to the anode on one side of the fuel cell, while oxidant (oxygen or air) is channeled to the cathode on the other side of the cell.

2. At the anode, a platinum catalyst causes the hydrogen to split into positive hydrogen ions (protons) and negatively charged electrons.

3. The polymer electrolyte membrane (PEM) allows only the positively charged ions to pass through it to the cathode. The negatively charged electrons must travel along an external circuit to the cathode, creating an electrical current.

4. At the cathode, the electrons and positively charged hydrogen ions combine with oxygen to form water, which flows out of the cell.

Backing layers
Hydrogen
Oxidant
Oxidant flow field
Hydrogen gas
Hydrogen flow field
Unused fuel
Water
Anode (negative)
Cathode (positive)
Polymer electrolyte membrane

SOURCE: Creative Commons.

5.2 Fuel cell vehicle developments

Auto-manufacturers are investing billions in developing fuel cell vehicles and even in launching the necessary infrastructure. The OEMs (auto-manufacturers) who are most behind fuel cell vehicles are Daimler, General Motors, Honda, Hyundai and Toyota. With the exception of General Motors, all four will have commercial fuel cell vehicles on the market for 2015. These OEMs have already built hydrogen-powered fuel cell cars, but the real challenge is to bring costs down. In the global race to do this, UK technologies are in a strong position, but they are not the only ones.

The US Secretary of Energy, Steven Chu, who once stated publicly that fuel cell cars weren't viable, has now come around to acknowledging their potential, especially in the light of the massive supply of natural gas from shale in the United States. In July 2012, the US Department of Energy (DOE) released its final report for a technology validation project that collected data from more than 180 fuel cell electric vehicles over six years. Although it is already out of date, it shows that their performance exceeded expectations, which is spurring a new round of research and development in America.

5.2.1 Bringing costs down

The fuel cell systems now under development are forecast to cost about $50 per kW. But the UK's Carbon Trust calculates that to compete with internal combustion engine cars, this cost must be reduced to about $35/kW. Achieving this could boost the market share of fuel cell vehicles by 10% by 2050. These extra vehicles equate to an additional $25bn of polymer fuel cell market value, saving 220 million tonnes of CO_2 globally. The overall market in 2050 could be worth $210bn, and save 750 million tonnes of CO_2.

To achieve this 30% cost reduction, significant breakthroughs are needed to reduce the cost, size and weight of a cell. Increasing power density is chief amongst these. The Trust is sufficiently convinced that membrane technology developed by ITM Power has the potential to achieve the $35/kW level that it has invested £1.1 million to develop and scale it up.

Reducing the amount of expensive platinum used by polymer fuel cells would reap further cost reductions. ACAL Energy has developed a revolutionary new design called FlowCath, inspired by the human lung and blood stream, which creates a low cost, virtually platinum-free system, and the Carbon Trust has invested £850,000 here, too. The Trust is also backing Southampton-based Ilika plc's development of a completely platinum-free catalyst with £150,000 of investment to support its commercialisation. Ilika's technology promises, on a cost/performance basis, to be 70% cheaper than the current industry standard. These investments come from the Carbon Trust's Polymer Fuel Cells Challenge (**http://www.carbontrust.com/client-services/technology/innovation/ polymer-fuel-cells-challenge**).

Daihatsu has gone further and developed a cheaper, completely platinum-free anion exchange fuel cell. This uses nickel or cobalt as a catalyst and hydrazine-hydrate as a fuel instead of hydrogen. Toyota has a majority stake in Daihatsu and the company is demonstrating concept cars such as the FC ShoCas, Kei, Pico and D-X that use this technology. The company is seeking partners to take this to commercial stage.

5.3 Hydrazine hydrate

The fuel for this novel fuel cell, hydrazine hydrate ($N_2H_4 \bullet H_2O$) is a non-flammable, high energy density fuel that can be stored in a technically safe manner. In an anion exchange membrane fuel cell, hydrazine breaks down to form nitrogen and hydrogen, which bonds with oxygen, releasing water. It can be produced in a low carbon process through the oxidation of ammonia using bleach, or hydrogen peroxide, to produce a zero carbon fuel that can be used in Daihatsu's anion exchange fuel cell vehicles. This is an exciting new technology seeking active development.

Previously used as a rocket fuel, like hydrogen, the liquid is an aqueous solution of hydrazine, and is produced by several well-established

FIGURE 5.3. Daihatsu's ShoCase concept van, which runs on hydrazine hydrate.

SOURCE: Daihatsu.

industrial methods. The fuel is liquid, easy to handle during filling and its energy density is high. It therefore has the huge advantage that it can potentially be distributed using the existing infrastructure for gasoline distribution.

5.4 Hydrogen vehicles

Maybe hydrazine hydrate is a game-changer. But at the moment, hydrogen is the most common fuel because platinum-containing proton exchange membrane fuel cells are the most common fuel cells. The vehicles powered by them emit no carbon emissions – only water – and offer significant environmental benefits, making them, like other electric vehicles, perfect for urban situations, especially in areas subject to tight separation controls. Hydrogen vehicles are 40–60% efficient compared to an internal combustion engine's efficiency of 25–30%, but the real test is an overall, 'well-to-wheel' (lifecycle) comparison of carbon emissions. This depends on how the hydrogen is generated.

Hydrogen and hydrazine-powered vehicles have two other advantages over other electric vehicles: they refuel as quickly as a conventional petrol car, and their range is at least twice as far: 300 km. This makes them ideal for long-distance driving. Hyundai's Dr Ing. Sae Hoon Kim believes that ultimately hydrogen fuel cell vehicles will play a strong role in what will inevitably be a mixed picture for personal transport.

Their main drawbacks are price and durability; the fuel cell stacks can only last for about 100,000 km or 7500 hours. Previously, there was a danger of the water that is output from the car freezing inside under certain conditions, but this has been eliminated and tested in Arctic conditions.

Hydrogen may also be burnt in a conventional internal combustion engine that has been modified in a similar fashion to LPG-burning engines. The result is a combustion process that is up to 25% more efficient than the use of gasoline. However, the engines are up to 1.5 times more expensive.

5.4.1 Hydrogen

Hydrogen, the simplest chemical element, reacts with oxygen, releasing heat that can be used in an engine or a fuel cell. The reaction also produces water, the only thing that comes out of the tailpipe. Although it sounds like a dream energy source, it has not yet reached mainstream because reforming it uses much energy, and its energy density per unit is much less than hydrocarbon fuels.

5.4.2 Hydrogen generation

Hydrogen is produced by separating it from water or methane, a process called 'reforming'. The currently preferred method is through the Steam Methane Reforming (SMR) of natural gas. This accounts for over 95% of production in the United States and 48% globally. At high temperature, steam is reacted with methane to produce carbon monoxide and hydrogen. This method emits more carbon dioxide than just using the methane (LPG) as a fuel, so will not help tackle global warming. Other methods for producing hydrogen are by using coal, other hydrocarbons or biomass as a feedstock, the recovery of hydrogen by-products from electrolytic processes, such as the manufacture of chlorine and ammonia, or from various refinery and chemical streams, and through the electrolysis of water.

It is the latter process, stimulated by renewable electricity, which is the green holy grail of sustainable transport. Electrolysis uses electricity to split water, H2O, back into its components: hydrogen and oxygen. Key to the widespread adoption of this process is decarbonisation of the electric grid, in order to enable surplus electricity, which cannot be used at the time of generation, to be used to make climate-friendly hydrogen. In other words, hydrogen is a perfect storage medium for renewable energy. (Sustainable methane, made from the anaerobic digestion of organic waste, could also be used to create climate-friendly hydrogen, although it is more efficient to use it directly as a fuel.)

In the United Arab Emirates, the production of hydrogen from methane has been proposed as a solution to climate change, with the carbon dioxide by-product being pumped back into oilfields for sequestration and to provide pressurisation to improve the extraction of oil. A £2 billion project to bring such a system online by 2014 was abandoned in 2011 after BP pulled out, blaming poor economic conditions. The Masdar Institute in Abu Dhabi is presently researching further uses of hydrogen in Arab states as a route to adaptation to a post-oil economy.

What is the global warming effect of driving a hydrogen fuel cell vehicle today, if the hydrogen is electrolysed using mains electricity in the UK? Assuming 56 kWh of electricity produces 1 kg of hydrogen, which is the claim of hydrogen production company ITM, then, to make it, 24.8 kg CO_2 is sent into the sky. This compares to 10.472 kg of carbon dioxide emitted on average for every gallon of petrol, which has about the same energy content (121 MJ) when burnt in a car. Accounting for the fact that the hydrogen vehicle is around twice as efficient as a petrol-driven car, this means that there is little difference between their emissions overall.

This is not an argument against hydrogen, because the grid is being decarbonised; by around 2020, when there could be a projected 3000 fuel cell cars on Britain's roads, 35% of Britain's electricity will be low carbon, on the road to 97% by 2030, as set out in the UK's 4th Carbon Budget, a document produced by the Committee on Climate Change and adopted by the government.

Besides, there are several prototype facilities in the UK already producing hydrogen for vehicles from renewable sources, and more are planned. A huge advantage for hydrogen is that any small-scale renewable electricity generator, like a wind turbine or solar array, can be set up right next to a refuelling station, producing hydrogen 24/7 for use close by. The wind blows at night and you can't use the electricity? Save it in hydrogen.

5.5 **Fuelling infrastructure**

For sustainable hydrogen vehicles to become widespread, alongside grid decarbonisation there needs to be created a fuelling infrastructure. Current thinking is that, if it happens at all, wide roll-out will not be reached until around the middle of this century.

Germany and South Korea took the early lead. Germany's programme of building 50 hydrogen stations by 2015, called H2Mobility, is going ahead, with a longer-term goal of 1000 stations by 2025. In 2011, Linde and Daimler set up a partnership to build 20 refuelling stations by 2014, and in 2012 Air Liquide announced it will build 10 stations. In the United Kingdom there is a similar initiative called UKH2Mobility. Wales has a programme of installing a series of refuelling stations along the M4 corridor that connects the Severn Bridge with Cardiff and Swansea. Denmark and France are also copying this model. In Japan, auto OEMs have teamed up with hydrogen and energy providers to plan a 100-station deployment in key metropolitan areas by 2015. With the exception of California, the US is falling behind.

The likelihood is that infrastructure roll-out will first occur in niche areas which are subject to tight emission control restrictions, such as London. The London Hydrogen Partnership (LHP) is a public-private partnership that has been working for some time to develop not just an infrastructure but advanced fuel cells and component manufacturing in the capital city. It is actively backed by the Greater London Assembly and the mayor's office and is trialling a number of technologies. It is a board member of AVERE – The European Association for Battery, Hybrid and Fuel Cell Electric Vehicles, a network promoting hydrogen fuel cells across Europe.

The UK government believes that the use of hydrogen as a fuel and energy storage medium has a significant future in the country. Therefore, the UKH2Mobility project, launched at the beginning of 2012, enrolled 13 industrial partners to co-evaluate potential roll-out scenarios for hydrogen for transport in the UK. More recently, five projects have been announced that will demonstrate the use of fuel cells and hydrogen and show how they can be integrated with other energy and transport components, such as renewable energy generation, refuelling infrastructure and vehicles.

Case study: ITM

ITM is a chief player in hydrogen generation and refuelling in the UK. It is managing Hydrogen On Site Trials (HOST) of its transportable high pressure refuelling unit (HFuel). Twenty-one commercial partners have taken part in 15 successful trials to date, which involve the loan of two Revolve HICE transit vehicles and one week's free use of the HFuel unit to power them. ITM Power has also partnered with Shell Research Limited and the National Grid plc in a study to investigate the technical, financial and operational feasibility of injecting hydrogen gas, generated from electrolysis fed from excess renewables, into the UK gas networks.

One of the five UKH2Mobility projects is a hydrogen refuelling station on the Isle of Wight. The island has a vision, like Taiwan's below, called EcoIsland. It aims to become completely renewable energy-powered by 2030. The Hydrogen Energy Storage Smart Grid has received a £2.4 million grant. £1.3 million of this has gone direct to partner ITM to build a 100 kg/day hydrogen refueller for a fleet of 20 vehicles, and find ways to use hydrogen as a storage medium to balance supply and demand.

5.5.1 Niche markets

In certain niche markets, fuel cell vehicles are already securing an advantage. Chief amongst these is in replacing battery-powered vehicles, because of the reduced charging time and longer range. Marks & Spencer, Coca-Cola, Walmart and FedEx have converted forklift trucks in their warehouses to fuel cells because they keep going for many times longer than battery-powered trucks. The fuel cell stack manufacturers have designed the fuel cells to be exactly the same size as the battery packs they are replacing, so the forklift trucks can be adapted quickly and simply. Customer reaction has been extremely positive. Such niche applications will accelerate the development of fuel cell vehicles generally.

A fleet of five fuel cell taxis running on hydrogen has been developed by UK power technology company Intelligent Energy, who provided the fuel cell systems for their power-train. They were assembled by the

FIGURE 5.5.1. London taxis filling up on hydrogen.

SOURCE: Intelligent Energy.

London Taxi Company. The taxis were used to ferry visiting guests of the Greater London Authority around London during the 2012 Olympic and Paralympic Games. A hydrogen fuelling station was opened in the summer of 2012 at Heathrow Airport as part of the developing London hydrogen network.

Built and managed by Air Products, it refuels the taxis as well as sometimes London's five-strong hydrogen bus fleet that usually operate on route RV1 from Covent Garden and Tower Gateway, and are part of the Mayor of London's drive to improve air quality in the capital. One fuel tank of hydrogen allows a hydrogen bus to run for at least 18 hours: 1000 fuellings enable the buses to travel approximately 100,000 miles around the capital, emitting only water from their exhaust pipes. Lea Interchange in Stratford, north London, which opened in 2010, is the location of what is currently the capital's only other hydrogen fuelling station, which serves the buses, which are operated by First. It can also serve privately owned hydrogen fuel cell vehicles, such as the Hyundai ix35 FCEV.

5.6 Electric fuel cell scooters

Scooters represent over two-thirds of road transportation in Taiwan. There are 1.4 million on its roads. Why is this important? Because it is in many islands throughout the world, not to mention urban areas, that the planet's 50 million scooters produced per year are to be found. Electrifying even a small proportion of these would be a huge market. Scooters are therefore in a strong position to catalyse change in places where other forms of transport, and air quality, are of low standard. Fuel cell hydrogen-powered scooters are on the way.

Just as Better Places has found a market in offering battery swaps in Israel (see above), Italian company Acta SpA is offering a hydrogen canister exchange service. Low pressure hydrogen canisters that fit into electric scooters are exchanged when depleted at special kiosks, via a vending machine that refills them using locally generated solar energy and an electrolyser. Back in Taiwan, the government has used a $300 million low carbon island development project on its Penghu Islands to trial electric scooters. This was the first large-scale eco-island project in the world. It is now deploying 80 fuel cell scooters and refuelling points in tourist locations in southern Taiwan. A large-scale project using Acta's refuelling kiosk system and electrolyser, which is of the alkaline membrane type, will begin to roll out in the next couple of years.

5.7 **Future prospects**

Currently, prices of hydrogen fuel cell vehicles are a rather steep five times greater than conventional cars, but these are mostly research and demonstration models. Prices will come down as technical innovations, production processes and mass production develop. Hydrogen fuel cell vehicles and electric vehicles complement each other, in that, because range is an issue, EVs tend to be smaller, lighter vehicles, while fuel cell cars are SUVs and sedans with twice the range. This helps auto-manufacturers to produce a range of low or zero-emission cars across their range of models.

Since 2000, Hyundai has produced 200 FCEV SUVs. The current model, ix35, is to go into production in 2013 with the first thousand rolling off the line to become part of test fleets around the globe. It will use an induction motor, not a permanent magnet motor, and have a 525 km range. It can

FIGURE 5.7. Hyundai's concept car, the ix35.

SOURCE: The author.

do 1–100 kmph in 12.5 seconds, with a 160 km/hr maximum speed. The vehicles are produced on the same line as conventional vehicles. The fuel cell is installed in the same place as the engine. Hyundai is intending to go into mass production with the cars in 2015, at which point about 11 other models are expected on the market from 11 auto-manufacturers.

The Toyota fuel cell hybrid vehicle on sale now features four hydrogen fuel tanks, an electric motor, a nickel-metal hydride battery and a power control unit. By 2015 it plans to have a zero-emission hydrogen-powered passenger car in forecourts. Toyota is supplying BMW AG with its own hybrid drivetrain systems and hydrogen fuel cell technology. This represents the first time Toyota has supplied its fuel cell technology to a rival carmaker, and follows an agreement the two companies cemented in December 2011. The advantage for Japan-based Toyota is that it will achieve greater economies of scale for its gasoline-electric hybrid systems and sport a trophy customer. In return, BMW is to supply diesel engines to Toyota, as well as its expertise on light, carbon fibre car bodies and the two companies will collaborate on the development of a lithium-ion battery.

Mercedes Benz, along with Daimler, is also building a fully automated fuel cell production facility in Burnaby, British Columbia, Canada.

....

Conclusion

BILLIONS OF POUNDS ARE BEING INVESTED in developing new sustainable fuels and the vehicles which will be powered by them. The impetus is coming from worldwide legislation concerning vehicle emissions and climate change. In the near term, biofuels occupy a much larger segment of the market than anything else. However, at least in the European Union, the recent decision to limit biofuels content to 5% of all fuels after 2020 will curtail the growth in first-generation biofuels that we have seen over the last few years. This means that there will be more focus on second- and third-generation biofuels, in commercialising them for wider deployment, particularly after 2020.

Once universally accepted sustainability criteria for biofuels are established, their role will continue its inexorable expansion as additives to conventional fuel. There are significant opportunities in promising second- and third-generation biofuels, for example jatropha and algae, which avoid competition over land for food crops. The need to reduce emissions from aviation creates a market for bio-kerosene that will drive their development. It will be some years before they become cost-competitive with conventional gasoline.

In addition, we will see many more models of electric vehicle, whether completely electric, hybrid or plug-in hybrid cars, scooters and vans, coming onto the market to increase consumer choice. Replacing the millions of scooters in use worldwide with electric versions represents a rewarding, niche application. Concurrent with this, battery technology is set to undergo a revolution, and any design which increases capacity and battery life will be a winner.

Key to all successful players will be the ability to form successful partnerships with other companies in the value chain, especially OEMs, as Ilika has with Shell and Toyota. There are significant opportunities for investments to be made in setting up the charging infrastructure for electric vehicles, and, to a lesser extent, in partnering with governments on creating a hydrogen refuelling infrastructure. Initially, business fleets are a prime target and driver of this market, as are motorists in inner city areas seeking small electric vehicles for short runs.

Within three to ten years, commercial fuel cell electric vehicles will appear and begin to make their presence felt in areas where hydrogen refuelling facilities are being made available with government support. However, it will be at least 15 years before they can compete in price with comparable conventional vehicles. The platinum-free anion exchange fuel cell, which runs on hydrazine hydrate, is also an interesting new technology that could make use potentially of the existing gasoline distribution network, but other proton membrane exchange fuel cell designs which reduce the use of platinum will find a large market. Hydrogen fuel cell vehicles complement battery-driven vehicles; the former for long distances, the latter for shorter ones.

For investors, this is a market with huge potential but still large risks. This overview of the present market and level of technology hopefully gives them a sound basis on which to evaluate potential winners.

....

Further Information

References in the text, in order of mention:

Legislation and general support

Institute for Transportation & Development Policy (ITDP): **www.itdp.org**

IEA's *Technology Roadmap: Fuel Economy of Road Vehicles*: **http://bit.ly/Sg4VVY**

IEA's *Policy Pathway: Improving the Fuel Economy of Road Vehicles*: **http://bit.ly/OlnYU2**

European Commission: European Green Cars Initiative: **http://bit.ly/VwvFFZ**

The Society of Motor Manufacturers and Traders (SMMT): **www.smmt.co.uk**

Transport & Environment: Clean vehicles portal: **http://bit.ly/QyLNUf**

Renewable Transport Fuel Obligation (RTFO) and related matters: **http://bit.ly/9qreNg**

Renewable Energy Directive: **http://bit.ly/S5I9UZ**

UK Committee on Climate Change: **theccc.org.uk**

US Renewable Fuel Standard II: **www.epa.gov/otaq/fuels/renewablefuels/**

US Bioenergy Program for Advanced Biofuels: **www.rurdev.usda.gov/BCP_Biofuels.html**

IEA Blue Map Scenario: www.iea.org/work/2011/egrd/day1/Remme.pdf

UK Department for Business, Innovation and Skills' Automotive Unit: www.bis.gov.uk/policies/business-sectors/automotive

UK Department for Transport's Office for Low Emission Vehicles (OLEV): www.dft.gov.uk/topics/sustainable/olev

UK Low Carbon Vehicle Partnership (LowCVP): www.lowcvp.org.uk

Cenex: www.cenex.co.uk

Biofuels

European Biofuels Technology Platform: www.biofuelstp.eu

European Commission's Strategy for a Sustainable European Bioeconomy: bit.ly/yiPYTn

Technology Roadmap: Biofuels for Transport (IEA, 2011): www.iea.org/papers/2011/biofuels_roadmap.pdf

Hydrocarbon Oil Duties Act 1979 (HODA): http://www.legislation.gov.uk/ukpga/1979/5

European Commission (Euractiv) report: bit.ly/NdBPLA

Bioenergy review (Committee on Climate Change, December 2011): www.theccc.org.uk/reports/bioenergy-review

EurActiv report, 30 January 2012, on EC review of the 1998 Fuel Quality Directive: http://bit.ly/Qj4kDL

Biofuelwatch: www.biofuelwatch.org.uk

US Department of Energy HDRD support: http://1.usa.gov/SaorJ2

IEA Bioenergy Implementing Agreement: www.ieabioenergy.com

Global Bioenergy Partnership (GBEP): **www.globalbioenergy.org**

Roundtable on Sustainable Biofuels (RSB): **rsb.epfl.ch**

International Organization for Standardization: **www.iso.org**

International Sustainability and Carbon Certification System (ISCC): **www.iscc-system.org**

Stockholm International Water Institute (SIWI): **www.siwi.org**

The study for the NGOs ActionAid and the RSPB of a proposed 50,000-hectare jatropha plantation in the Dakatcha woodlands of Kenya: **http://bit.ly/SZrmpi**

Lux Research biofuels reports: **http://bit.ly/RqEhK2**

Pruning the Cost of Bio-Based Materials and Chemicals: **http://bit.ly/UhHGfY**

UK Technology Strategy Board: **www.innovateuk.org**

EVs

International Transport Forum: **www.internationaltransportforum.org**

International Council for Clean Transportation: **www.theicct.org**

Electric Cars: Ready for Prime Time? (International Transport Forum, July 2012): **http://bit.ly/KxJGRv**

Low Carbon Vehicle Demonstrator: **http://bit.ly/ikEMp**

Battery technology charges ahead McKinsey (July 2012): **http://bit.ly/L6sE8O**

UK Department of Transport, 'Plugged-in Places': **http://bit.ly/U9jelw**

US Department of Energy EG Project: **http://1.usa.gov/PPpbBB**

UK 'Plugged-in Fleets Initiative': http://bit.ly/xEcori

Smart Grids and Electric Vehicles: Made for each other? (OECD's International Transport Forum, July 2012): http://bit.ly/N7bMi4

Association of European Automotive and Industrial Battery Manufacturers: www.eurobat.org

Fuel cells

US Department of Energy (DOE) fuel cell report: http://1.usa.gov/SIOm9O

UK Carbon Trust: www.carbontrust.com

Germany's H2Mobility project: http://bit.ly/OCRPqV

UKH2Mobility: http://bit.ly/AiALMY

The London Hydrogen Partnership (LHP): www.london.gov.uk/lhp/

AVERE – The European Association for Battery, Hybrid and Fuel Cell Electric Vehicles: www.avere.org

Hydrogen On Site Trials (HOST): http://bit.ly/PokPPS

Other

Energy conversion factors: http://bit.ly/OaV5tz

International Council on Clean Transportation: www.theicct.org

Acknowledgements

THANKS TO THE FOLLOWING PEOPLE who have helped with research and fact-checking: Jo Abbess, Helen Adam, Simon Crawford, Frank Jackson, Tim Pullen.

About The Author

DAVID THORPE writes environmental material, non-fiction, fiction and journalism. He has been the News Editor of Energy and Environmental Management magazine since 2000. His books include *Solar Photovoltaics Business Briefing* (Dō Sustainability, 2012), the *Earthscan Expert Guide to Sustainable Home Renovation* (Earthscan, 2010) and the *Earthscan Expert Guide to Solar Technology* (Earthscan, 2011). He also writes or has written for Greenpeace, Grant Thornton, *Business Green*, Corlan Hafren (a group comprising engineering consultancy Halcrow, Arup, and KPMG), the *Guardian Online* and many others. He was Managing Editor at Centre for Alternative Technology Publications in the 1990s. His non-fiction includes an educational book for kids, *How The World Works*. He is also the author of *Hybrids* (*'Essential reading for the cyberspace generation'*), winner of the HarperCollins-Saga Magazine 2006 Children's Novelist competition, and has written and edited many scripts, comics and cartoon strips. He was a co-founder of the London Screenwriters Workshop, and co-wrote *The Fastest Forward* for Comic Relief, a feature film starring Jerry Hall. His career includes being the only person in the world (probably) to hold a degree in Dada and Surrealism! He also finds time to play in a band and run a media company, Cyberium. He lives in Wales and was born in Robin Hood country, Nottingham.

For Product Safety Concerns and Information please contact our EU
representative GPSR@taylorandfrancis.com
Taylor & Francis Verlag GmbH, Kaufingerstraße 24, 80331 München, Germany

www.ingramcontent.com/pod-product-compliance
Ingram Content Group UK Ltd.
Pitfield, Milton Keynes, MK11 3LW, UK
UKHW040928180425
457613UK00011B/305